Christianity
and the Nature of
Science

Christianity

and the Nature of

Science

J. P. Moreland

BAKER BOOK HOUSE
Grand Rapids, Michigan 49516

ISBN: 0-8010-6249-7

Fourth printing, April 1994

Printed in the United States of America

Library of Congress Cataloging-in-Publication Data

Moreland, James Porter, 1948–
     Christianity and the nature of science.

     Bibliography: p.
     1. Religion and science—1946—   . 2. Science—
Philosophy. I. Title.
BL240.2.M645   1989   261.5'5   89-6719
ISBN 0-8010-6249-7

To

Bob Blair, Bill Roth, Walt Russell, Dallas Willard

These are among the Dunedain
Men of Old, Men of Valor

Contents

Acknowledgments

Several people played special roles in my life while I was writing this book. They helped me, in various ways, to complete it. First, several professors at Liberty University have encouraged me, though of course the views expressed are mine and do not necessarily meet their approval—Professors David Beck and Gary Habermas (philosophy); John Morrison and Dan Mitchell (theology); Dick Patterson (Old Testament); Robert Littlejohn, Paul Sattler, and Bruce Triplehorn (biology); and Robert Chasnov (physics).

Second, I wish to thank my students at Liberty University and Grace Fellowship Church in Baltimore. They endured my sometimes pedantic and exploratory wanderings in science, philosophy, and theology. Richard Loper in particular made helpful comments on a rough draft.

Third, a group of serious Christian lay persons provided finances that allowed me to buy materials for study and to have leisure for reflection. Their actions set a standard for other Christians who wish to promote Christian scholarship. Jim and Jeannie Duncan, Tim and Bobbi Smick, Ben and Laura Abel, and Jonathan and Suzanne Glenn all befriended me in this and other ways.

Fourth, several of my friends inspired and sustained me through their lives and examples: Walter Bradley, Don Smedley, Marty Russell, Patty Roth, Scott and Sally Rae, Klaus and Beth Issler, Bill Craig, and Norman Geisler. I also want to mention E. Calvin Beisner. His philosophical and editorial comments on an earlier draft of the manuscript were invaluable.

Finally, my wife, Hope, and our daughters Ashley and Allison provided family identity and unity that gave me strength for the hard work and long hours of writing.

Introduction

It is no mere truism to say that we live in the modern world. People tend, at least in the Western world, to think in secular, naturalistic terms. True, cults and other deviant forms of religious expression are on the rise. But these phenomena underscore the secular nature of the modern world, they do not refute it, because they are perceived by those in the modern *Zeitgeist* as irrational, emotional, even romantic expressions of the human thirst for meaning, not as rationally defensible conceptualizations of the religious mode of life.

The Christian community is called to witness to and interface with modern culture in a humble, Christ-honoring, and well-informed way. Over seventy-five years ago evangelical Presbyterian scholar J. Gresham Machen wrote, "False ideas are the greatest obstacles to the gospel. We may preach with all the fervor of a reformer and yet succeed only in winning a straggler here and there, if we permit the whole collective thought of the nation or of the world to be controlled by ideas which, by the resistless force of logic, prevent Christianity from being regarded as anything more than a harmless delusion." Machen's penetrating insight is as true today as ever. Theologian R. C. Sproul has called this the most anti-intellectual period in the history of the church. Charles Malik, former president of the United Nations, warns that the greatest danger facing modern evangelicalism is anti-intellectualism.

This brings us back to the modern world. Undoubtedly the most important influence shaping the modern world is science. People who lived during the Civil War had more in common with Abraham than with us. From space travel and nuclear power, anesthesia and organ transplants, computers and brain bisections, to DNA research, optics, and lasers, ours is a world of modern science.

If the church is to speak to the modern world and interact with it responsibly, it must interact with modern science. Christians cannot afford to promulgate a dichotomized stereotype of Christianity wherein a believer's spiritual life is a private, individualized faith operating in some upper story (to borrow Francis Schaeffer's term) while his secular life is public and involves reason and argument.

11

The purpose of this book is to assist and encourage Christians to think more clearly about the relationship between science and theology. There are different models for how science and theology should be integrated. I discussed several in *Scaling the Secular City* and will not pursue that discussion in detail here. Hence it may be helpful to say a word about a particular model of scientific and theological interaction: the complementarity view.

According to this view, science and theology are complementary, noninteracting, noncompeting descriptions of the world. They use different languages and methods and ask and answer different questions. Thus, the phenomenon of conversion can be described from a chemical, biological, psychological, sociological, or theological point of view. Each vantage point is complete in its own sphere and uses its own language. None of them, even in principle, interacts or conflicts with any other. Applied to the question of the origin and development of life, the complementarity view takes theistic evolution as the proper way to integrate theology and science.

I respect this approach to integration. It offers much to anyone thinking about science and religion, and there are several areas where science and religion do complement each other; for example, several facets of conversion may be described psychologically or sociologically in a way that complements theological descriptions. Advocates of this approach deserve thanks for their hard thinking and the rich resources they have given to the church to aid in the integrative task.

However, the complementarity view fails as a total model of integration. It can and should be incorporated into a total picture of the science/theology interaction, but it is the tail that must not wag the dog. The complementarity view is too clean, too neat, too easy in its summary dismissal of areas of science and religion that, prima facie, appear to conflict. If taken as an entire model, the complementarity view tends to compartmentalize science and religion. It also tends toward fideism. Thus it fails to respond responsibly to secular questions about why one should believe in God in the first place, especially granted how those questions are framed in a scientistic culture. Further, it comes perilously close to deism in its reduction of much of God's primary causal activity (direct, immediate, miraculous interventions into the natural world) to secondary causality (the indirect, mediate use of natural processes to secure some divine intent). It also fails to give proper value to areas where science is incomplete. When those areas are predictable from a normal, grammatical/historical hermeneutic of Scripture, they present legitimate opportunities for believers to point beyond the physical world to the spiritual and are not mere god-of-the-gaps invocations to cover our ignorance.

Holy Scripture, particularly when it speaks of the origin and de-velopment of life and of man's dual nature, should not be understood along the lines of the complementarity approach. There are areas where science and theology make claims at the same level of description. These claims do compete with each other, and they occasionally con-flict. In what follows, therefore, I assume that the complementarity view is inadequate and that science and theology really do interact on common ground.

How, then, should we solve tensions between science and theology? Are not science and religion like oil and water—they don't mix? Is not creation science an attempt by narrow-minded fundamentalists to foist what belongs in Sunday school on students who want to study science? Isn't science an autonomous set of disciplines that, concep-tually speaking, have little to do with theology?

The purpose of this book is to discuss these and related questions in order to help believers see that science and theology have interacted with each other and should. In order to defend such a claim, the dis-cussion incorporates insights of philosophy, especially philosophy of science, because how science and theology should interact is primarily a *philosophical* question (though science and theology are involved).

More specifically, I attempt to defend three theses:

First, there is no definition of science, no set of necessary and suf-ficient conditions for something to count as science, no such thing as *the* scientific method, that can be used to draw a line of demarcation between science and nonscience. Nothing about science essentially excludes philosophical or theological concepts from entering into its very fabric. Science is not an airtight compartment isolated from other fields of study, and there is nothing unscientific about creation science.

Second, limits to science arise in a number of interesting ways, and these limits are sufficient to do two things: 1) They show that scien-tism—the view that science alone is a rational approach to the world that secures truth—is false. 2) They weaken the epistemic authority of science, depriving it of its claim to dominate or overrule theology and philosophy. The interaction between science and theology or phi-losophy is a dialogue, not a monologue.

Third, attempts to integrate science and theology, including efforts to resolve apparent conflicts between them, should not automatically assume a view of science known as scientific realism. Scientific real-ism, roughly defined, is the view that successful scientific theories are true or approximately true models of the theory-independent world. A number of antirealist approaches to science agree that science works—it solves problems, gives us predictions, allows us to control nature and describe observations simply—but that its success does not indicate that scientific theories are true or approximately true. An eclectic model of science, one that uses a realist or antirealist view of science on a case-by-case basis, should be used to integrate science and theology.

Chapter 1 discusses attempts to define science and state a set of necessary and sufficient conditions for something to be called science. It also discusses various ways that philosophy interacts with science. Chapter 2 discusses scientific methodology and concludes that there is no such thing as the scientific method. Rather, science uses a family of methodologies, and various aspects of these methodologies are used in disciplines outside science. The chapter ends with an example of how theology uses scientific methodology.

Chapter 3 discusses various types of limits to science and argues that scientism is naive and false. Chapter 4 defines scientific realism and considers various objections to it. Chapter 5 describes a number of antirealist views of science and concludes with a brief sketch of an eclectic model of science that could be used to integrate science and theology. Finally, chapter 6 considers the claim that creation science is not science but religion in disguise. The chapter defends the scientific status of creation science by offering a working characterization of what it is supposed to be and by responding to criticisms of its status as a science.

I have offered a large number of resources in the notes (in fact, a larger number than one usually finds in philosophical works) in order to lead readers to sources for further study. I hope this book, plus the resources cited, will stimulate and motivate my Christian brothers and sisters to do more work in the philosophy of science, regardless of whether they agree with me.

My own intellectual background includes a degree in chemistry, a graduate degree in theology, and two graduate degrees in philosophy. I am barely a layman in science and do not consider myself primarily a theologian. My primary academic vocation is that of a philosopher, but even here, as anyone who studies the philosophy of science can attest, it is hard to come to terms with all facets of the philosophy of science since it is such a detailed, sometimes mind-boggling area of study.

Thus, while I recognize my limitations, I offer this work as a serious attempt to explore areas of interaction among science, philosophy, and theology. If the arguments seem persuasive, well and good. Wherever you disagree, I invite you to develop alternatives.

The Christian community can only benefit from such an exchange. Our community is mature enough to embrace wide differences of opinion in these areas. But we greatly need more work on science and theology that is sensitive to both the history and the philosophy of science. If this book stimulates such work, I will consider it successful. It is written primarily for Christians, either as a textbook in formal schools or informal church or parachurch educational activities, or as a guide to the philosophy of science for believers who want to integrate science and theology. It can also be given to unbelievers who are interested in why Christians think theological concepts are relevant to science.

The Definition of Science 1

Philosophy is the only rational knowledge by which both science
and nature can be judged. By reducing philosophy to pure science
man has . . . abdicated his right to judge nature. . . .
—Etienne Gilson, *The Unity of Philosophical Experience*

The victory in the Arkansas [creation science] case was hollow,
for it was achieved only at the expense of perpetuating and can-
onizing a false stereotype of what science is and how it works.
—Larry Laudan, "Science at the Bar—Causes for Concern"

From the beginning of the Christian church to the present, Christians have felt it necessary to integrate their theological doctrines with
rational beliefs from scientific, philosophical, and other sources. According to the great theologian Saint Augustine (354–430), "We must
show our Scriptures not to be in conflict with whatever [our critics]
can demonstrate about the nature of things from reliable sources. . . ."[1]
Since the believer takes Christianity to be true and rational, and since
Christianity makes claims about the way the world is and how it came
to be, then in principle it is possible for competing claims to conflict
with biblical revelation. Thus, the church has historically found that
Augustine's advice is part of her mission in the world.

Today, things have changed. For a number of reasons (e.g., Kant's
critique of religion and the reduction of theology and revelation to
anthropology and religion), believers and unbelievers alike have come
to understand religion in general and Christianity in particular such
that they cannot, even in principle, conflict with scientific or philosophical claims about the world. Supposedly, something about the

1. Augustine *De genesi ad litteram* 1.21. Cited in Ernan McMullin, "How Should
Cosmology Relate to Theology?" in *The Sciences and Theology in the Twentieth Century*,
ed. Arthur R. Peacocke (Notre Dame: University of Notre Dame Press, 1981), p. 20.

very nature of religion isolates it from other disciplines of study, especially science.

The essence of religion, we are told, is to help people's private, practical, moral lives, to offer meaning and purpose in life. It does not matter if a religion is true, especially when it steps out of its boundaries and makes ontological or etiological claims; what matters is that it work. By contrast, it is supposedly part of the very nature of science that it gives us true and rational information about the world and how it came to be. Thus religion and science, by their very nature, mix as well as oil and water—not at all. The major factors contributing to this cultural myth have been the purported warfare between science and religion, the dazzling success and progress of science, and the alleged naturalism inherent in scientific methodology and scientific laws and theories.

Nowhere has this understanding of the definitions of science and religion been more evident than in the recent controversies about creation science. We are told repeatedly in newspapers, magazines, television talk shows, and scholarly publications that science rules out religion by definition. According to Robert C. Cowen, the natural science editor for the *Christian Science Monitor*, "It is this many-faceted on-going science story [the theory of evolution] that should be told in public school biology courses. Creationists want those courses to include the possibility of—and 'scientific' evidence for—a creator as well. There is no such 'scientific' evidence. The concept of a supernatural creator is inherently religious. It has no place in a science class."[2] In other words, the very definition of science rules out any interface with religion that, as Augustine would have seen it, involves a prima facie conflict and thus requires an attempt to integrate the two into a coherent synthesis.

According to this widespread understanding of the scientific enterprise, science is a tightly circumscribed set of disciplines that, by its very nature, excludes theology, philosophy, and other "nonscientific" fields of investigation. These fields are not relevant to the methodology and practice of science. Neither are they sources of information about the world that should be consulted in assessing some particular truth claim in a given scientific law or theory.

In this chapter we will investigate the claim that science is a distinct, isolated approach to the world wherein claims from other fields, especially philosophy, are irrelevant to scientific methodology and theories. (We will leave to later chapters a discussion of theology and its

2. Robert C. Cowen, "Science Is What Can Be Argued, Not What Is Believed," *The Baltimore Sun*, July 8, 1987.

integration with science.) We will investigate various definitions of science that seek to distinguish between science and nonscience. We will look at philosophy and show how it relates to science. But first, let us take a brief look at definition in general.

The Act of Definition

Why do people define things and seek definitions?[3] Usually a definition is sought in order to eliminate ambiguity and clarify meaning, since the same word can be used in a variety of contexts with different meanings. But sometimes a word is defined to influence attitudes in a (frequently) question-begging, honorific, or degrading way. C. S. Lewis once remarked that the term *medieval* had become a term of abuse. If something is "medieval" it is outdated and somewhat pedantic. The term *scientific* is often used in an honorific sense; that is, if something is "scientific" it is good, rational, modern, and if something is not "scientific" it is old fashioned and not something a fully actualized person will believe.

This honorific use of *scientific* often occurs in discussions about science and Christianity, but we will not examine it further. It is propagandistic. It works by creating positive associations and imagery (after all, who wants to be "unscientific" in this sense?) that cloud issues of substance and bypass rational argument.

More important for our purposes is the use of definition to clarify meaning and eliminate ambiguity. This use is often associated with a definition by "genus and difference," that is, an attempt to state the essential attributes of the thing defined, attributes that are identical in all examples of that thing. For example, Aristotle defined man as a rational animal. Every example of a human (e.g., Socrates, Alexander, and Plato) is a rational animal. Rational animality is, in this definition, the essence of being human.

Two things should be pointed out about this sort of definition. First, one does not need to know the definition of something before he can recognize clear cases of the thing being defined. Terms like *love, justice,* and *history* are virtually impossible to define, but I can recognize cases where my wife, Hope, is showing love to me, examples where justice

3. For a useful introductory discussion of definition, see Irving M. Copi, *Introduction to Logic*, 3d ed. (London: Macmillan, 1969), pp. 89–120. Elsewhere, Copi has discussed the important distinction between real and nominal essences as it relates to definition in "Essence and Accident," reprinted in *Naming, Necessity, and Natural Kinds*, ed. Stephen P. Schwartz (Ithaca, N.Y.: Cornell University Press, 1977), pp. 176–91.

is aborted, and instances where someone engages in historical research. Similarly, a chemist titrating an acid with a base is doing science, while a fortuneteller consulting a crystal ball is not. Clear cases like these can be recognized without a definition of science. Definitions are most helpful, but in the greatest danger of being question begging, in borderline cases. We will return to this observation about definition, clear cases, and borderline cases later in the chapter.

Second, because it is often virtually impossible to state the essence of something one is trying to define, it is usually wiser to state necessary and sufficient conditions for the application of the term in question.[4] Here terms are "defined" according to the following scheme:

P, if and only if, Q

Suppose you were trying to define what it was for something to be triangular. P would be "x is triangular." Q could be replaced with "x is trilateral" (i.e., "x has three sides"). In other words, even though trilaterality does not really amount to the same property as triangularity, nevertheless, if some object under investigation is to count as having three angles, it is a necessary and sufficient condition for that object that it be trilateral.

In defining science, we will examine various attempts to state necessary and sufficient conditions for something to count as science. If such conditions can be stated, then we will have grounds for drawing a clear line of demarcation between science and nonscience.

Finally, questions about the nature of definition in general and attempts to define science in particular are philosophical issues, not scientific issues per se.[5] For the question *What is the proper definition*

4. Two terms can be materially equivalent or coextensional, that is, one term is true in all and only the cases where the other term is true. This relationship is weaker than the identity relationship. For a treatment of material equivalence and identity as they are used in metaphysics, see Michael Loux, *Substance and Attribute* (Dordrecht, Holland: D. Reidel, 1978), pp. 77–85.

5. Some philosophers have denied that there is any real difference between philosophy and science, claiming that philosophical issues are really just scientific issues broadly considered. See W. V. O. Quine, *Ontological Relativity and Other Essays* (New York: Columbia University Press, 1969), pp. 69–90. For a brief statement and critical response to Quine, see Jonathan Dancy, *Introduction to Contemporary Epistemology* (New York: Basil Blackwell, 1985), pp. 233–39. For an overview of different ways philosophers see the relationship between philosophy and science, especially regarding the attempt to characterize science as a discipline, see Ernan McMullin, "Alternative Approaches to the Philosophy of Science," in *Scientific Knowledge: Basic Issues in the Philosophy of Science*, ed. Janet Kourany (Belmont, Calif.: Wadsworth, 1987), pp. 3–19.

of science? is itself a philosophical question about science that assumes a vantage point above science; it is not a question of science. One may need to reflect on specific episodes in the history of science to answer the question. But the question and the reflection required to answer it are philosophical in nature, a point not diminished merely because a scientist may try to define science. When she does so, she is doing philosophy.

Defining Science

Let us begin our search for a definition of science by listing some representative proposals:

John J. O'Dwyer in *College Physics:* "... [science] seeks to understand the world of reality in terms of basic general principles ... involving observation, intuition, experimentation, debate, and reformulation."[6]

William Keeton in *Biological Science:* "Science is concerned with the material universe, seeking to discover facts about it and to fit those facts into conceptual schemes, called theories or laws, that will clarify the relations among them. Science must therefore begin with observations of objects or events in the physical universe."[7]

Webster's New Collegiate Dictionary: "3a: Knowledge covering general truths or the operations of general laws esp. as obtained and tested through scientific method b: such knowledge concerned with the physical world and its phenomena."[8]

None of these definitions is adequate. The first and second are offered by scientists and illustrate that scientists today, in contrast to their counterparts in earlier generations, are often ill-equipped to define science, since such a project is philosophical in nature. Consider the first definition. Dwyer begs the question in favor of a realist understanding of science by stating that science seeks to "understand the world of reality." We will see in chapters 4 and 5 that a number of scientists and philosophers do not think that science necessarily at-

6. John J. O'Dwyer, *College Physics*, 2d ed., (Belmont, Calif.: Wadsworth, 1984), p. 2.
7. William T. Keeton, *Biological Science*, 3d ed. (New York: W. W. Norton, 1980), p. 3.
8. *Webster's New Collegiate Dictionary*, 1975 ed., s. v. "science."

tempts to understand reality. Much of what we would want to count as science involved scientists who merely tried to "save the phenomena," that is, to provide ways to harmonize, collate, and predict sensory experiences of the world without believing that they were describing the hidden structure of the world of reality.[9] Second, scientists do not always explain things by general principles. Dwyer seems to assume a *covering-law* model of scientific explanation whereby scientists explain particular facts (this copper wire expands when heated) by subsuming them under general laws (all metal expands when heated; this copper wire is a metal). But he is wrong. A covering-law model of explanation often is used outside of science, and within science it is not necessary. Scientists often explain things by using models or pictures (e.g., treating atoms as billiard balls). They also find it necessary on occasion to explain things by postulating a cosmic singularity, some brute particular state of affairs taken as a given. Postulation of the Big Bang is an example. The conditions in the Big Bang are postulated in an attempt to explain the unfolding of the universe, but the Big Bang is not itself a general principle. Finally, though it is not clear, Dwyer seems to equate reality with what scientists investigate. If so, he clearly begs the question since values, numbers, and so forth may be a part of reality that scientists do not investigate.

What about the second definition? It is also inadequate. For one thing, it is not clear that science investigates only the material universe. Biology, psychology, anthropology, paleontology, and sociology investigate living things, their relationships, their remains, and their artifacts. But it is not obvious that these are merely physical, nor is it clear that scientists investigate only physical aspects of them. To cite one example, psychologists investigate the structures of consciousness and subconsciousness, but a good case can be made that these features of human beings cannot be reduced to material phenomena. Second, one could substitute the word *philosophy* in Keeton's definition for the word *science*. Among the things philosophy attempts is the discovery and conceptualization of facts about the material universe. There may be a difference in how scientists and philosophers go about their work, and there may even be a difference in their aims, but Keeton's definition does not even begin to alert the reader to such issues. Further, some philosophers believe that philosophy starts with observation, in some sense at least. For example, Thomas Aquinas's (1225–1274?) proofs for God's existence begin with observations of the existence of finite, dependent beings. Finally, we will see in chapter 2

---

9. This is illustrated nicely by Rom Harre, *The Philosophies of Science: An Introductory Survey* (Oxford: Oxford University Press, 1972), pp. 62–99.

that science almost never begins with observation. Rather, the path to scientific discovery often, though not always, begins with the formulation of a question followed by a tentative guess about a hypothesis that guides the scientist in knowing what is relevant to observe.

The third definition is somewhat better. But as we have already seen, it is wrong to identify the object of scientific knowledge with general truths or laws (historical sciences like geology and some aspects of astronomy focus on particular events, e.g., the extinction of the dinosaurs). But the upshot of this definition is that science must be defined in terms of what is called the scientific method. We will investigate scientific methodology in the next chapter, so we need not comment on this aspect of a definition of science here.

Testing a Court's Definition of Science

None of the definitions of science so far considered has been adequate. Another attempt to define science is the late Judge William R. Overton's listing of allegedly essential features of science in his decision against scientific creationism in the famous creation science trial in Little Rock, Arkansas, in December 1981. He wrote: "More precisely, the essential characteristics of science are: 1) It is guided by natural law; 2) It has to be explanatory by reference to natural law; 3) It is testable against the empirical world; 4) Its conclusions are tentative, i.e., are not necessarily the final word; and 5) It is falsifiable."[10] Let us examine each of these in turn. We begin by combining points 1 and 2 and rewording them to make them even stronger.

NATURALISM AND NATURAL LAW

Something is scientific if and only if it focuses on the natural world, is guided by natural law, and/or explains by reference to natural law. Let us break this down into three different claims and consider first the phrase *guided by natural law.* What does this mean? If it means "seeks to explain in terms of natural law" then this second condition reduces to the third one ("explains by reference to natural law"), and we will consider this later. Perhaps it means "motivated by a desire to find a natural explanation." If so, then this is clearly not a necessary or sufficient condition for something to count as science. For example, some philosophers who are not theists may be motivated to find a natural

10. Cited in Norman L. Geisler, *The Creator in the Courtroom* (Milford, Mich.: Mott Media, 1982), p. 176.

(i.e., nonsupernatural) explanation of the existence and nature of morality, but they are not doing science.

In contrast, a large number of men in the history of science did science from a motivation to please God and think his thoughts after him. Further, some of them have practiced science with the belief that no natural explanation of a particular phenomenon was available. For example, the great botanist Carl Linnaeus (1707–1778) founded modern taxonomy in 1735 with his work *Systema Naturae*. Linnaeus, a creationist, was motivated and guided by his belief that no natural explanation was available for the existence and nature of living organisms. Nevertheless, his work was clearly an example of science.

It should be clear that people can have a variety of motivations for carrying on their work, and their private motivation is not particularly relevant in assessing the question of whether their activities and theories are scientific.

What about the phrase *explains by reference to natural law?* Is this a necessary or sufficient condition for something to count as science? No. Mathematicians often explain things by referring to natural (nonsupernatural) laws of mathematics or logic. In fact, some mathematicians and philosophers believe that these laws really exist and take them to be relations that can tie together abstract entities called propositions that can be in people's minds. Moral philosophers sometimes explain why some act is wrong, for example, murder, by appealing to some natural moral law—for instance, that people should not be treated only as means to an end but as ends in themselves. Edmund Husserl (1859–1958) studied the natural law of foundation, for example, the relation between the way a red color varies in some whole as the size of the whole diminishes, or the way a whole cannot have an instance of color without that whole first being extended in space. On the basis of studies like these, Husserl went on to formulate several natural laws about part–whole relations that explain various phenomena like the ones just mentioned. When mathematicians and philosophers behave in this way, they are explaining things by reference to natural laws they believe exist, but they are not doing science.

Conversely, scientists do not always engage in explaining by reference to natural law. We will see in chapter 2 that there are other models of scientific explanation that do not involve explaining by appeal to a natural, scientific law.

Further, scientists sometimes explain something by appealing to a brute given that is not itself a scientific law and is not capable of being subsumed under more general laws. As mentioned earlier, the Big Bang is one example; others would be the various physical constants (e.g., the rest mass of an electron). The whole notion of an ultimate particle, which some scientists clearly use, is another. Perhaps we have not discovered the ultimate particle yet, perhaps we will have to go through several further candidates until we do, but if and when we do, the existence and values for the properties of that particle (e.g., its rest mass, charge, and so forth) will be brute givens incapable of explanation by reference to natural law. The point here is that the notion of an ultimate particle, even if there are no ultimate particles, is not unscientific. It is similar to the notion of a first cause, a brute given, or a cosmic constant that is just there and cannot be explained by reference to a scientific law.

Finally, for a long time scientists and philosophers have recognized the distinction between establishing the existence of a phenomenon and explaining it. As Larry Laudan puts it:

> Galileo and Newton took themselves to have established the existence of gravitational phenomena, long before anyone was able to give a causal or explanatory account of gravitation. Darwin took himself to have established the existence of natural selection almost a half-century before geneticists were able to lay out the laws of heredity on which natural selection depended. If we took the *McLean* Opinion [Judge Overton's] criterion seriously, we should have to say that Newton and Darwin were unscientific; and to take an example from our own time, it would follow that plate tectonics is unscientific because we have not yet identified the laws of physics and chemistry which account for the dynamics of crustal motion.[11]

Now someone may object to some of the arguments listed in the following way: "You have cited mathematicians and moral philosophers as examples of explaining by reference to 'natural' law." But surely this is a different sense of *natural* from the one scientists use when they use the word *natural*.

This objection provides a transition to the third part of the essential characteristic of science we are considering—something is scientific if and only if it *focuses on the natural world*. There are two different ways we could take the term *natural*. First, it may be a contrast term that means *nonsupernatural*. Here "natural" means everything that is real apart from God and his direct, primary causal interventions into the world. Naturalism would be the view that God does not exist and everything that does exist is a natural entity. Or a theist could adopt an attitude of naturalism by simply focusing on items insofar as they exist in the world, even though she believes that God created them,

11. Larry Laudan, "Commentary: Science at the Bar—Causes for Concern," *Science, Technology, & Human Values* 7 (Fall 1982): 18.

that is, she could focus on them from a natural point of view.[12] Whether atheistic or theistic, naturalism so understood could embrace, in addition to carbon atoms, electromagnetic waves, and procaryotic cells, such entities as numbers, Cartesian souls, moral values, sets, laws of logic, universals (entities like triangularity, wisdom, humanness, and redness that are capable of being in more than one thing at the same time), and private visual sense data. These entities are not likely to be physical, but they are natural entities as defined above.

If this is how we should take *natural*, then science is not the only discipline that "focuses on the natural world." The entities listed are studied by philosophers, logicians, mathematicians, and linguists, to name a few. Even theologians study and make claims about natural entities, for example, by saying that God created various forms of life, endowed them with certain abilities, and so forth. So if *natural* is understood in this way, it is clear that science cannot be defined as the discipline that "focuses on the natural world."

But perhaps a second understanding of *natural* is appropriate for distinguishing science from nonscience. *Natural world* can be taken to mean the world of physical things having only physical properties that are part of one spatiotemporal system.[13] This would rule out the various nonphysical entities listed two paragraphs above. Is this definition adequate for defining science? Again, no. For one thing, psychologists, sociologists, anthropologists, paleontologists, histologists (histology is a branch of anatomy that focuses on tissues), and biologists do not clearly study physical things. Some psychological theories, like those of Freud and Jung, explain behavior by appealing to conscious or unconscious desires, intentions, and other mental states. Sociologists study, among other things, cultural systems, roles and relationships, and cultural artifacts (e.g., the relationship between art and social class), but none of these objects of study is clearly physical. Consider a painting. Someone can attend to it at the level of a physical thing, a heap of markings on a page. But at that level it is not art. Only by focusing on aesthetic qualities and patterns that emerge in

12. For examples of Christian scientists who argue that science must adopt the naturalistic attitude as I have operationally defined it, see Richard Bube, *The Human Quest: A New Look at Science and the Christian Faith* (Waco: Word, 1971), pp. 17–49; "The Nature of Science," in *The Encounter Between Christianity and Science*, ed. Richard Bube (Grand Rapids: Eerdmans, 1968), pp. 17–42; David A. Young, "Is 'Creation-Science' Science or Religion?—A Response," *Journal of the American Scientific Affiliation* 36 (September 1984): 156–58; Paul de Vries, "Naturalism in the Natural Sciences: A Christian Perspective," *Christian Scholar's Review* 15 (1986): 388–396.

13. Cf. D. M. Armstrong, *Universals and Scientific Realism I: Nominalism and Realism* (Cambridge: Cambridge University Press, 1978), pp. 26–32.

the painting taken as an aesthetic whole, on entities that are not clearly physical, does one get at the painting qua a work of art. Only then is the object of investigation a cultural artifact. Biologists study ecosystems, animal behavior, animal perception, and so forth. But these are not clearly physical.

Now, someone could argue that, in fact, all the psychological, sociological, and biological items I have listed are really physical after all. But such a response would miss the point. These items *may not* be physical. Whether they are is a philosophical question, and psychologists' current practice of studying them as if they were physical will not somehow become unscientific if such entities are not physical. These current practices are rightly deemed scientific irrespective of what answer to the philosophical issue wins the day. But if, by definition, science only studies physical things, and if some philosopher comes up with a convincing argument that human desires, animal sensory states, and the like are not physical, then suddenly we would have to treat the current scientific study of these things as nonscientific. In fact, some people in the history of science (scientists with views akin to those of George Berkeley [1685–1753] and Ernst Mach [1838–1916]), including some contemporary physicists (e.g., some who study quantum physics) do not believe that matter exists, but their studies of space, time, and so forth surely count as science.[14]

Finally, philosophers, as well as scientists, study the material world.

14. For an overview of the philosophy of science held to by Berkeley and Mach, see John Losee, *An Historical Introduction to the Philosophy of Science*, 2d ed. (Oxford: Oxford University Press, 1980), pp. 159–65. For an overview of eight different ways of understanding quantum physics, some of which affirm the nonexistence of a mind-independent material world in general or of quantum phenomena in particular, see Nick Herbert, *Quantum Reality* (Garden City, N.Y.: Doubleday, 1985), pp. 15–29. Now all of these approaches to quantum physics are consistent with scientific practice, but since some of them deny a mind-independent material world, then science, even physics, must be capable of being identified without settling the philosophical question of the existence and nature of matter. For philosophical aspects of the problem of defining matter, see Howard Robinson, *Matter and Sense: A Critique of Contemporary Materialism* (Cambridge: Cambridge University Press, 1982), pp. 108–23; Ernan McMullin, ed., *The Concept of Matter in Greek and Medieval Philosophy* (Notre Dame: University of Notre Dame Press, 1963); *The Concept of Matter in Modern Philosophy* (Notre Dame: University of Notre Dame Press, 1963). Science itself has changed its views on what matter is. Did Newton fail to practice science because we no longer believe in the sort of matter he visualized? Obviously not. One could argue that we count as scientific someone who focuses on an entity as material whatever his conception of matter and whatever later discoveries show matter to be. But if such an argument allows us to count Newton as a scientist, why could it not also be used to count Berkeley and Mach? The fact is that the concept of matter has its own problems of vagueness; thus its appropriation in a definition of science is not as straightforward as is often thought, even for the harder sciences.

Their methods may differ from those of science and their questions may differ, but that is another issue. The fact is that philosophers ask questions about the nature of space and time, whether there are ultimate particles and, if so, whether they are like very small billiard balls or like some sort of continuous stuff like molasses (to use Bertrand Russell's analogy) or a field, whether the properties of physical things are themselves physical, and so on.

In sum, we have not been able to show clearly how this first alleged essential characteristic of science can be used to draw a line between science and nonscience. Let us look at Judge Overton's criterion.

EMPIRICAL TESTABILITY

Something is scientific if and only if it can be tested against the empirical world. This statement brings out two different issues—testing and observation. We will look at testing and falsification shortly. Here we will focus on observation. With this in mind, the criterion under consideration means something like this: Something is scientific if and only if it appeals to observational data to judge whether it is true or reasonable to believe. In chapter 4 we will investigate in more detail the nature of observation, the observation/theory distinction, and related issues. But for now, three points should be made.

First, it often happens that two scientific theories are empirically equivalent.[15] That is, they have all and only the same observational consequences, and one cannot choose between them without appealing to something besides observational considerations, for example, theoretical simplicity or internal clarity. In this case, the two scientific theories cannot be tested against one another by observation, but must be judged by an appeal to philosophical, epistemic virtues good theories presumably should have. Some theories are empirically equivalent, not in principle, but only for a while. For example, various

15. Actually, the issue of empirically equivalent theories enters into discussions at three levels. First, there are empirically equivalent philosophical views about the nature of the external world and sense perception, e.g., the debate between John Locke (we directly see our ideas of the world, which are caused by the external world), George Berkeley (we directly see our ideas of the world, but God causes them and no intelligible sense can be given to a mind-independent or unperceived world), and Thomas Reid (we see objects in the world—e.g., chairs and dogs—directly, we do not see our sensory ideas of those objects). Second, some claim that different philosophical views about science are empirically equivalent, e.g., scientific realism, operationalism, phenomenalism, pragmatism. More will be said about this in chapter 5. Third, some scientific theories themselves are empirically equivalent, either in principle or for large stretches of time when no testable observations are available. Debates of this last kind include those between Copernican and Ptolemaic astronomers (1540–1600), wave and particle optics (1820–1850), atomists and anti-atomists (1815–1880).

competing theories (some empirically equivalent in principle) about celestial motion were offered by Eudoxus (c. 400–c. 347 B.C.), Ptolemy (fl. A.D. 150), Copernicus (1473–1543), Tycho Brahe (1546–1601), and Kepler (1571–1630).[16] Some theories are empirically equivalent in principle, so no observation whatever could decide between them. The late J. L. Mackie has offered an argument to show that, contrary to the special theory of relativity, there is an absolute, preferred space-time frame of reference within which notions of absolute rest and motion make sense.[17] The details of Mackie's arguments need not concern us, but his theory conflicts with the Special Theory of Relativity, the theories are empirically equivalent, and both are scientific.

Second, the observation/theory distinction has been criticized. As a result, many philosophers and scientists no longer believe that theory-independent observation is possible. So there may not be any such thing as a scientific theory confronting the empirical world by means of a naive, unmediated observation of "the given." And even if there is such a confrontation, precious few of the observations we count as scientific are examples of it.

For one thing, scientific theories often deal with unobservable entities and processes such as quarks and electromagnetic fields. Thus, these entities must be tested indirectly by observation and there is an uneliminable theoretical component to the scientific meaning of those entities that cannot be tested by observation. The total content of these theoretical concepts is not exhausted by their observational aspects, which are themselves not entirely unstained by theoretical interpretation.

Further, seeing something usually involves *seeing as* or *seeing that*. Seeing an electron, for example, involves seeing something *as* an electron or seeing *that* it is an electron. Seeing *as* and seeing *that* are not mere passive receptions of sensory input. Rather, they involve describing the electron as a theoretical entity with such-and-such properties. That is, seeing *as* and seeing *that* involve a degree of interpretation in

16. See Harre, *Philosophies of Science*, pp. 66–89; Thomas Kuhn, *The Copernican Revolution* (Cambridge: Harvard University Press, 1957).

17. J. L. Mackie, "Three Steps Towards Absolutism," in *Space, Time, and Causality*, ed. Richard Swinburne (Dordrecht, Holland: D. Reidel, 1983), pp. 3–22. While Mackie's case rests on philosophical argument, he does suggest that his views make sense if we adopt the Lorentz-Fitzgerald contraction hypothesis as a solution to the null results obtained in the famous Michelson-Morley experiment. For a good description of that experiment, see Rom Harre, *Great Scientific Experiments* (Oxford: Oxford University Press, 1983), pp. 114–24. For a survey of the various attempts to explain the Michelson-Morley results, including an account of the Lorentz-Fitzgerald solution, see Robert Resnick and David Halliday, *Basic Concepts in Relativity and Early Quantum Theory*, 2d ed. (New York: John Wiley and Sons, 1972), pp. 12–29.

the very act of seeing. For example, prior to the 1900s, when someone saw the sun, he was seeing atoms that (assuming the correctness of current radioactive decay theory) were breaking down into smaller parts or fusing into larger ones. But no one knew that, since the relevant theory was not in place. So no one saw the sun *as* a center of fission or fusion reactions, no one saw *that* this was the case. Scientific seeing is normally, perhaps always, seeing *as* or seeing *that*. Neither of these involves a naive form of simple observation often assumed in attempts to define science, for example, a straightforward testing against an unmediated, empirical world.

Finally, Peter Achinstein has argued that what counts as an observation is so context-dependent that in certain cases an entity counts as being observed while in other cases it counts as an unobserved theoretical entity postulated to explain what is observed. Thus, there is no hard-and-fast sense that can be given to the idea that some scientific datum can be described in exclusively observational language.[18] For example, in some contexts one can claim to observe an electron (I saw the electron leave a track through the cloud chamber), but not in others (electrons must move through the wire to explain the current registered on our meter).

In sum, the observation/theory distinction is sufficiently vague to warn us against appropriating a naive view of observation in an attempt to define science.

Third, disciplines other than science appeal to observation. Historians appeal to observation of historical facts when they attempt to substantiate their claims. So do scholars involved in literary studies. For example, suppose someone were trying to decide how to interpret a particular word used by Plato in one of his dialogues. A literary scholar could appeal to observations about the use of that word in similar contexts elsewhere as a part of his case.

Philosophers sometimes appeal to sensory forms of observation as well. For example, some philosophers, like Husserl, who defend the existence of universals—entities like redness, which can be in more

18. Peter Achinstein, *Concepts of Science: A Philosophical Analysis* (Baltimore: Johns Hopkins, 1968), pp. 157–78. Achinstein offers a painstaking analysis of the plethora of uses for the term *to observe*. He uses "seeing" and "observing" in different senses. Thus, he argues that seeing, unlike observing, does not require attending to aspects or features of something, while observing does. (I can see a man walk quickly by my door without observing him, if I fail to notice him by not fixing my attention on him.) I have not bothered to distinguish seeing from observing. For more on seeing, seeing as, seeing that, and perception in general, see Frank Jackson, *Perception: A Representative Theory* (Cambridge: Cambridge University Press, 1977). Jackson's otherwise excellent book is, unfortunately, mired in a representative dualist view of perception in the tradition of John Locke.

than one thing at the same time—against nominalists, who deny their existence, have pointed out that one can simply observe that two different balls have literally the same color in them. Philosophers also appeal to a broader form of sensation or intuition in discussing propositions like "If P, then Q, and P, therefore Q" and "Something cannot be both red and green all over at the same time."[19] Some have argued that we know these propositions are true by means of an immediate, rational perception of the propositions themselves. Theologians also appeal to observation: religious experiences (called numinous perception)[20] of perceiving God, experiences of seeing life *as* created by God (or *seeing that* it is a creation of God), seeing that the events surrounding Israel's regathering are acts of God, and so forth.

In sum, there are several genuine problems with any attempt to use observation as a necessary or sufficient condition for something to count as science.

TENTATIVE BELIEF

Something is scientific if and only if it is held tentatively.

This criterion is almost totally wide of the mark. For one thing,

19. Knowledge of synthetic a priori propositions, at least those that are not like the identities used in science, e.g., water is H_2O, involve "observation" in two ways (and more typically scientific identities probably do as well): sensory modes of intuition (e.g., seeing several red and green objects) as occasions for synthetic a priori knowledge, and rational modes of intuition (e.g., simply seeing immediately that something cannot be red and green all over at the same time) as immediate justifications for such knowledge. See Paul K. Moser, ed., *A Priori Knowledge* (Oxford: Oxford University Press, 1987), especially the article by Roderick Chisholm. Admittedly, this latter form of "observation" is not exactly the same as that intended in a good bit of scientific practice. But two things should be kept in mind. First, as has already been pointed out in the preceding note, the use of "observation" in science lies on a multifaceted continuum that sometimes merges with more rational forms of intuition (seeing a theoretical entity like a quark involves a good bit of conceptual interpretation and rational intuition into background theory and its internal conceptual relationships, not unlike seeing that a proof is correct in logic or mathematics). Second, rational intuition is foundational for more sensory modes of observation; thus science presupposes the former. For more on this last point, see George Bealer, "The Philosophical Limits of Scientific Essentialism," in *Philosophical Perspectives, Vol. I: Metaphysics, 1987*, ed. James E. Tomberlin (Atascadero, Calif.: Ridgeview, 1987), pp. 289–365; less technical and very valuable in this regard is Richard M. Zaner, *The Way of Phenomenology: Criticism as a Philosophical Discipline* (Indianapolis: Bobbs-Merrill, 1970), pp. 41–78, especially pp. 51–62.

20. For a comparison of sensory and numinous perception, see J. P. Moreland, *Scaling the Secular City: A Defense of Christianity* (Grand Rapids: Baker, 1987), pp. 234–40. For more on the role of observation in forming and testing the truth claims of theological propositions, see Douglas Macintosh, *Theology as an Empirical Science* (New York: Macmillan, 1919), pp. 1–46; Holmes Rolston III, *Science and Religion: A Critical Survey* (Philadelphia: Temple University Press, 1987), pp. 1–22.

scientists do not always hold to their beliefs tentatively, especially during periods called normal science. Normal science, in contrast to a period of scientific crisis where a scientific theory or paradigm has a number of troublesome anomalies, is characterized by a period in which the scientific community is not willing to give up a core theoretical belief. Was Newton tentative about his belief in the existence of forces? Would any contemporary scientist seriously question the theory that blood circulates? Laudan, Thomas Kuhn, Imre Lakatos and a number of others have shown that dogmatism characterizes periods of normal science and that such an attitude often, though not always, has been very constructive in promoting the aims of science.

Further, philosophers, other scholars in the humanities, and theologians are often tentative about their views. For example, Christian theologians are often tentative, that is, open to new evidence, about a number of issues ranging from interpretations of specific passages to the inerrancy of the Bible and the existence of God. A distinction may be helpful at this point. Roderick Chisholm has shown that there are at least two different senses that can be given to the right to be sure of something and, thus, not to be tentative.[21] By the right to be sure one can mean that she has terminated inquiry and is no longer open to further evidence, or one can mean that she has the right to trust some item of belief and rely on it in forming and testing other judgments.

Most theologians use the second sense of "the right to be sure," and they hold this right with varying degrees of conviction depending on the issue and the evidence at hand. For example, if the Bible is in fact the Word of God, then it is the last word on something if it is properly interpreted. But one can still be open to further refuting evidence for this claim or for some particular interpretation of the Bible.

So it is just not the case that something is scientific if and only if it is held tentatively. Besides, claims about the relative merits of dogmatism and tentativeness are often associated with sociologically esteemed values embraced in a pluralistic society and can easily degenerate into ad hominem attacks. This should warn us against using the criterion under consideration in attempts to define science. It is liable to generate far more heat than light.

FALSIFIABILITY

Something is scientific if and only if it is falsifiable.
This line of demarcation between science and nonscience is usually

---

21. Roderick Chisholm, *Theory of Knowledge*, 2d ed. (Englewood Cliffs, N.J.: Prentice-Hall, 1977), pp. 116–18.

associated with the philosopher Karl Popper.[22] The role of falsification in science is not clear. As Rom Harre has said, "There is a widespread myth that scientists do experiments to test hypotheses."[23] In other words, the idea that scientists usually perform crucial experiments in order to falsify one of two or more rival hypotheses is not the way scientists normally practice science.

Nevertheless, falsification is certainly relevant to science. Whether it constitutes a necessary or sufficient condition for science, however, is quite another matter. We will discuss the role of falsification in the confirmation of scientific theories and laws in chapter 2. For now, three things should be mentioned.

First, the nature of falsifiability in science is often difficult to clarify. For example, seldom if ever are individual scientific propositions tested in isolation from other propositions or theories.[24] For instance, suppose a scientist predicted that if a photon were to be beamed on an atomic nucleus, then it would produce a pair of particles called an electron and a positron with certain properties. Now suppose that his observations do not square with his predictions. Something has been falsified. But what? A number of theories and auxiliary assumptions have been tested jointly. For example, our experiment will only work given certain assumptions about light, about the structure of an atomic nucleus, about the behavior of mass and energy, about the instruments we are using, and so forth. Let H stand for his hypothesis about particle production, and let C_i—C_n be the various auxiliary assumptions involved. Then these are related to the experimental observations O in the following way:

$$(H \,\&\, C_i \,\&\, C_j \,\&\, \dots C_n) \rightarrow O$$
$$\frac{\text{Not–O}}{\text{Therefore, Not–}(H \,\&\, C_i \,\&\, C_j \,\&\, \dots C_n)}$$

One cannot specify conclusively which hypothesis in the whole cluster of hypotheses has been falsified.

Again, this is not to say that falsification is not somehow relevant to science. But H and C_i—C_n, taken individually, are clearly scientific propositions even though they are not falsifiable in isolation from other scientific propositions. Similarly, the statement *God made man* may not be falsifiable by itself. But when it is embedded in an entire theory

22. Karl Popper, *Conjectures and Refutations: The Growth of Scientific Knowledge* (New York: Harper and Row, 1963), pp. 33–59.

23. Harre, *Great Scientific Experiments*, p. 108.

24. See Pierre Duhem, "Physical Theory and Experiment," reprinted in *Scientific Knowledge*, ed. Kourany, pp. 158–69.

that has falsifiable consequences, then it can be considered falsifiable as a member of a group of statements.

Furthermore, it is not always easy to specify when ad hoc hypotheses—that is, adjustments of a hypothesis made to save it from falsification—are inappropriate. And the attempt to state criteria for appropriate/inappropriate uses of ad hoc hypotheses is itself a philosophical question, not a strictly scientific one. The facts that criteria for appropriate uses of ad hoc hypotheses are somewhat vague, that forming such criteria involves philosophy, and that hypotheses are falsified in groups all combine to make it difficult to clarify falsification with the precision necessary to use it as a line of demarcation for science.

Second, science includes a wide range of empirical generalizations ranging from fairly specific to quite broad. At the former end of the spectrum would be, for example, some specific experimental test at a specific time for some specific property of an electron. More general would be the Bohr model of the atom (including Niels Bohr's [1885–1962] conception of an electron). More general still would be the corpuscularian version of atomism, roughly the view that ultimately there is only one type of substance, a Newtonian-type atom that neither comes to be nor perishes, and all change and secondary qualities (color, smell, taste, sound, texture) can be explained in terms of or reduced to the movements and collisions of atoms. More general still would be a commitment to atomism in general, that is, the commitment to the view that there are ultimate particles (even though we may not have found them yet) and that these are more like tiny billiard balls (not necessarily Newtonian corpuscles) than like a field (they are discrete). More general still would be some form of physicalism which would be a research program standing against vitalism, any form of mind-body dualism (except perhaps epiphenomenalism), and so forth.

Now the point in listing this ascending order of scientific generalities is to show that the farther one goes up the scale toward the more general end, the harder it is to falsify the relevant scientific theory. In fact, some scientific research programs like physicalism or a commitment to absolutism regarding space and time are virtually impossible to falsify, for one can always readjust them, often in appropriate ways, to avoid falsification.

Put another way, broad scientific research programs approach the status of general world views. Now, world views can be falsified in principle, at least some of them can (if the resurrection of Jesus is falsified then Christian theism is false), but doing so is very difficult because their epistemic support is so multifaceted. Broad research programs in science are like this as well, and they are not unscientific for that reason. It would be arbitrary at best to limit the definition of science to include only very low-level generalizations.

Third, disciplines outside science use falsification. Historians allow their hypotheses to be falsified by historical facts. Literary scholars allow their interpretations of a text to be falsified by word studies and the like. Theological claims about the acts of God in history, for example, the resurrection of Jesus of Nazareth, are open to falsification as well. As Holmes Rolston has pointed out, theological claims come in a spectrum of degrees of generality as well. And various religious claims, like "The family that prays together stays together," "God becomes dear in times of adversity," or "Praying with such-and-such attitudes tends to achieve greater results than praying with other attitudes" are testable and can be falsified.[25] So science is not the only discipline that uses falsification.

In addition to the criteria for science offered by Judge Overton, several others are widely believed to constitute conditions for counting something as science. Three of these are analyzed below.

Popular Misconceptions of Science

Something is scientific if and only if it is measurable or quantifiable.

In discussion about religion and science, often it is said that what marks off science from other fields is that it is quantifiable. Scientists can measure and quantify their data by applying mathematics to their investigations and theorizing.

But is it really true that something is scientific if and only if it is quantifiable? In spite of popular opinion to the contrary, no. For one thing, several scientific theories and areas of study do not involve quantification as an essential ingredient. For example, virus theories about disease transmission, various psychological models of the causes and cures of neuroses, most theories of paleontology, several aspects of the general theory of evolution, sociological theories about family systems and group interaction, theories in cultural anthropology, and so on, are not quantifiable.

In fact, the clearest case where quantification is important is in

25. Rolston, *Science and Religion*, pp. 7–8. Only the first example is his. See also Colin A. Russell, *Cross-Currents: Interactions Between Science and Faith* (Grand Rapids: Eerdmans, 1985), pp. 54–79, especially pp. 67–71. It might be objected that theologians often save their theories by using inappropriate ad hoc hypotheses. But this is a hard charge to prove and, in any case, we have already mentioned that it is hard to specify when ad hoc hypotheses are inappropriate even in science. For more on this, see Mario Bunge, *Method, Model, and Matter* (Dordrecht, Holland: D. Reidel, 1973), pp. 27–43.

most aspects of physics and chemistry. But that is simply because these fields focus mostly on matter as a quantifiable stuff. That is, much of the point of physics and chemistry lies in their viewing matter in a mathematical way. Other branches of science view their objects of study from other points of view, and they often focus on physical or nonphysical properties (e.g., properties of a virus or subconscious desires of a person) and patterns of relationships. But the fact that their approach often differs from that of physics and chemistry does not alone disqualify these branches of study from counting as science. The major point of a scientific theory, even in physics and chemistry, is usually not quantification but explanation, and the former is not always relevant to the latter.[26]

Furthermore, various disciplines outside of science use quantification in investigation. For example, in literature some scholars use statistical analyses of word frequencies to determine the authorship of a particular work. Statistical word studies can also help clarify an ambiguous word in a certain context by showing how many times the word is used, what percentage of cases in certain contexts have one meaning versus another, and so on.

Again, historians can use numerical data. A historian could argue that in a high percentage of cases of major cultural change economic factors were decisive and, therefore, in some case under consideration it is likely that economic factors were decisive as well. Even philosophers have quantified their data from time to time. Some of the early utilitarians, like Jeremy Bentham (1748–1832), tried to quantify the amounts of utility produced by different moral acts in order to determine a utilitarian calculus for moral decision making. Bentham failed, but he did use quantification, even though his efforts were philosophical, not scientific. In sum, it is neither necessary nor sufficient that an activity quantify its data for it to count as science.

Something is scientific if and only if it involves predicting new test results.

The role of prediction in science has occasioned considerable debate among philosophers, and one's view of how prediction relates to a scientific theory depends, in large measure, on one's philosophical theory about the nature and function of scientific laws and theories in

26. For examples of this, see Frederick Suppe, ed., *The Structure of Scientific Theories* (Urbana, Ill.: University of Illinois Press, 1977), pp. 62–66; Harre, *Philosophies of Science*, pp. 168–83. For a Thomist interpretation of the role of quantification in natural science, see James A. Weisheipl, "The Relationship of Medieval Natural Philosophy to Modern Science: The Contribution of Thomas Aquinas to its Understanding," *Manuscripta* 20 (1976): 181–96.

general.[27] Since different views about the nature and function of scientific laws and theories will be discussed in later chapters, we need not mention them here. For now, three points will show that prediction is neither a necessary nor a sufficient condition for something to count as science.

First, scientific theories do not necessarily use prediction. More fundamental to a theory is that it explain facts. For example, different scientific theories about the extinction of the dinosaurs do not necessarily predict new results, but they still count as scientific explanations. Even a theory in physics need not predict something to count as science.

To see this, consider an example. Suppose scientist A is trying to formulate a law describing the expansion of a metal as a function of heat. Suppose further that he has 50 experimental values or points on a graph for ordered pairs of length of expansion and amount of heat added. Suppose now that he correctly predicts a new point on his curve from a law he has formed from the 50 values he already has. He now has 51 experimental confirmations for his law.

Now consider scientist B. He has 51 points on the same curve and he retrodictively formulates the same law as scientist A. Scientist B has the same law and the same experimental confirmation as scientist A, but B did not predict something, while A did. In fact, it would be possible for there to be only 51 total points and no new ones forthcoming. This would be the case, for example, in theorizing about extinct animals like dinosaurs. It would be possible, for the number of fossils relating dinosaur death to climate to be finite and fixed. In that case, scientist A could formulate a law before all the fossils had come in and predict correlations for future digs, while scientist B could wait until all digs were complete to formulate his law.

The same point could be made about two scientists studying 51 orbits of some planet that suddenly blew up. Scientist A devises his law of the planet's motion after 50 orbits and predicts points on orbit 51. Scientist B retrodicts the same law after 51 orbits. If both laws were the same and if both covered or explained the data equally, then the presence or absence of prediction would be irrelevant to the question whether they were scientific. The past or future disappearance of some phenomenon, making prediction of new data impossible, does

27. W. F. Bynum, E. J. Browne, and Roy Porter, *Dictionary of the History of Science* (Princeton: Princeton University Press, 1985), s.v. "prediction." Positivists view the role of prediction in science as an attempt to control nature, instrumentalists (E. Mach) as the summarizing and predicting of sense data, deductivists (C. G. Hempel) as symmetrical to explanation and, thus, eliminable, inductivists (R. Carnap) as the acid test of a theory, realists as often present but secondary to understanding and explanation.

not remove the phenomenon under investigation or the laws/theories used to explain it from the domain of science.

Epistemic or logical relations that obtain between a law or theory and the facts it seeks to explain are not temporal in nature. There is an emotional commitment to the prediction of new data, because if it is verified it is surprising and exciting. New predictions can also help insure (though not guarantee) that a given law or theory was not formulated in an inappropriately ad hoc manner. Nevertheless, the main feature of a scientific theory is explanation, not prediction, and explanation involves logical, analogical, and epistemic relationships that are not temporal.

Second, some theories were not scientific despite the fact that they predicted new data. Consider an example.[28] The ancient Babylonians and Greeks differed in their methods of astronomy. Scientists in both cultures attempted to construct tables to predict various astronomical facts, such as, the rising and setting of the constellations, sun, and moon, throughout the year. The Babylonians constructed their table, which involved adding and subtracting constants according to specific calculating rules, by merely formulating the rules based on repeated observation of the locations of heavenly bodies. These devices allowed the Babylonians to predict astronomical phenomena accurately, but they did not explain why the heavenly bodies moved as they do.

Greek astronomers constructed models that allowed for the requisite predictions as well, but their models postulated explanatory entities, for example, spheres upon which the earth, moon, and so forth rested and by means of which they rotated. The Greeks and Babylonians both predicted phenomena accurately, but most thinkers have taken the Greek astronomers as better examples of scientific theorizing. Why? Because they explained facts and did not merely predict them. So prediction is not a sufficient feature for a theory to count as good science.

Finally, disciplines outside science use prediction as well. Historians predict that future events will obtain given certain circumstances

28. Harre, *Philosophies of Science*, pp. 45–47. Dudley Shapere argues that cases of ordered domains in science (like the periodic table, where items are classified into some pattern on the basis of some ordering rule or principle) do not become scientific theories merely because they allow for predictions. He cited the periodic table and the Hertzsprung-Russell diagram (an ordered domain in astronomy that plots the absolute magnitude of stars, i.e., the brightness stars would have if they were a standard distance from the earth, as a function of their temperature). Both were used to make predictions, but neither on the basis of any theory of what was going on. See Dudley Shapere, "Scientific Theories and Their Domains," in *Structure of Scientific Theories*, ed. Suppe, pp. 518–65, especially pp. 534–39.

or that new data will continue to verify certain explanations of a historical period after further research, and these predictions often are based on historical generalizations.[29]

Literary scholars use prediction as well. For example, a Pauline scholar could theorize, perhaps due to investigation of Paul's earlier epistles, that Paul uses the word *salvation* for legal justification in his polemical writings and for temporal deliverance from danger in his pastoral writings (this is only an illustration and not an attempt to represent Pauline usage accurately). This scholar could now predict that if certain features in one of Paul's later epistles, for instance, Philippians, indicate that it is pastoral, then one should find the temporal deliverance usage of "salvation."

Theologians also can make predictions. For example, one could predict that if certain spiritual disciplines are practiced, then certain religious experiences or certain patterns in one's family system will follow. Again, someone could predict that if religious awakening takes place, then certain results based on New Testament teaching should follow. One could then study past or present examples of religious awakening to see if the results tended to obtain.

It might be argued that the theological examples are defective, different in essential ways from scientific predictions. "Perhaps," one could argue, "the theological predictions do not use theories, are vague and general in nature, and do not make reference to sensory experiences as do their scientific counterparts." But the objection fails. For one thing, theological predictions do use theological models involving conceptualizations of God, the nature of spiritual life, and so forth, derived from both special and general revelation. Further, theological predictions involve a wide range of generality, as do their scientific counterparts. We have already discussed the distinction between low-level empirical generalizations and broad research programs in science. Predictions vary in generality from these accordingly. Very specific, testable predictions can be obtained for a specific test of a specific low-level generalization. But the same can be said for theological predictions. Someone can tell whether some specific prayer request, asked according to a specific passage of Scripture, was answered affirmatively. More general claims, like commitment to physicalism in science or to the providential care of God in theology, involve more general predictions that are harder to test decisively on a specific occasion.

Further, scientific predictions often are based on unseen theoretical entities that imply observational consequences. Theological claims do

---

29. See W. H. Dray, *Philosophy of History* (Englewood Cliffs, N.J.: Prentice-Hall, 1964).

sometimes involve numinous perception, for example, when one predicts that fasting and contrition of heart will make one's perception of God's presence deeper and richer. But even these types of predictions can be used to formulate and test lawlike generalizations about the development of the spiritual life.[30] And not all theological predictions appeal to numinous perception. The prediction that certain ways of praying tend to be answered affirmatively, that Christ will return, or that dispensational theology predicts that Israel will be preserved as a nation can all be tested by more normal modes of observation. So it is not the case that science and science alone makes predictions.[31]

Something is scientific if and only if it is repeatable.

Finally, some people believe that a necessary and/or sufficient condition for science is that the thing in question—fact, law, theory—be repeatable. However, there has been some confusion about what it means to say that something is repeatable. Obviously, since time does not move backward, no particular event can ever be strictly repeated.[32] What sense of repetition is meant here? At least two senses can be discerned.

First, something may be repeatable in an epistemological test, that is, in the sense that others can duplicate your procedures and see if the results obtain as you claimed they did. In other words, something is repeatable in this sense if it is publicly checkable and not sanctioned by a mere appeal to one's own personal, private, subjective awareness that cannot be shared by others. This sense of repeatability seems to be the one intended by Douglas Futuyma. Writing of the various trustworthy methods employed by scientists, he says, "What makes them trustworthy is repeatability. An observation is accepted as a scientific 'fact' only if it can be repeated by other individuals who follow the same methods."[33]

Now this does seem to be a necessary condition for something to

30. A good example of this can be found in Evelyn Underhill, *Mysticism* (New York: Meridian, 1955), pp. 167–451.

31. See Rolston, *Science and Religion*, pp. 4–8, 26–31.

32. Thus, scientific repeatability seems to presuppose the real existence of universals that can be literally repeated, and these are what scientists focus on. They often consider particular entities, e.g., an electron, not qua particular but qua instance of a kind of thing, an electron in general. For more on universals, see J. P. Moreland, *Universals, Qualities, and Quality-Instances* (Lanham, Md.: University Press of America, 1985). For an application of this point to the defense of biblical inerrancy, see J. P. Moreland, "The Rationality of Belief in Inerrancy," *Trinity Journal* 7 (Spring 1986): 75–86.

33. Douglas Futuyma, *Science on Trial: The Case for Evolution* (New York: Pantheon, 1982), p. 166.

count as scientific.[34] Scientists do make claims that they take to be checkable by others. The problem is that almost every other field of investigation that uses argument and appeals to the experience of the knowing subject does so as well. Arguments from religious experience appeal to a repeatable test for the presence of God. If one meets certain conditions—has a searching, contrite heart, prays to Jesus for salvation, and so forth—then one can test for herself that God will become real to her and change her life in fairly standard ways that can be largely specified in advance.

In literary or historical investigation, one can justify an interpretation of the main purpose of some text of the resurrection of Jesus by citing evidences, procedures, and explanations that one went through to arrive at or validate one's interpretation. An appeal can be made for someone else to repeat these steps for himself and test the result. The same claim can be made for mathematics, logic, philosophical argumentation, rational insight into the meaning and truth of certain synthetic a priori truths (e.g., redness is a color), and so forth. So repeatability in its epistemological sense is not something that marks off science from nonscience.

But there is a second sense that can be intended by repeatability, namely, an ontological sense. Here, the claim is made that something can happen over and over again, or at least that different instances of the same phenomenon can be repeated in reality (e.g., the sun rises each day), and that somehow this repetition is essential for science.

There are some examples of this in science. The clearest case is in the formation of statistical laws (Mendel's laws of genetics, or perhaps some statement to the effect that out of 1000 examples of a certain diseased animal, 90% are cured if given a certain medicine). Here, repetition is inherent in the very form of the statistical generalization since it involves reference to some frequency of events out of a class of events.

But not all scientific laws are of this type. Some laws are formed by conjecture and guessing. They are not codified generalizations derived statistically from enumerative induction. The Greek astronomers cited are an example of this. They postulated a universe of spheres,

34. Even this may be wrong. If one adopted a phenomenological approach to science like Berkeley's, then this view, and other logical positivist variants of it, could be faulted for being a form of solipsism. In that case, one could practice science in an irreducibly private way. I do not want to press this point since I think Berkeley was wrong, but the claim that science is repeatable in the sense of being public does seem to rest on a philosophical assumption that solipsistic varieties of Berkeleyan phenomenalism are false. So, even here, philosophical argumentation is in some sense prior to scientific characterization.

relying in part on metaphysical frameworks about the various regions of the universe. Their laws could be tested even if the phenomena they explained were not ongoing. Other examples are models about the origin of the universe, historical geology, paleontology, and related fields. Scientists form and test models about unique and unrepeated events in the past, and these models are scientific even though they do not seek to explain ongoing processes. Nor do they strictly use, in their explanations, laws or theories that are themselves mere statistical summaries of repeatable, ongoing events and processes.

It should be clear as well that fields outside of science may use statistical laws. A historian could appeal to a statistical law about the factors usually present in the fall of a culture to make probable an explanation of why some culture under investigation was decaying. Again, a proponent of the design argument for God could use a statistical law to the effect that such-and-such a percentage of highly ordered entities are designed by a mind, and go on to argue that the universe or some aspect of it like the eye or the human mind was most likely designed, too.[35] So neither the epistemological nor the ontological sense of repeatability draws a line of demarcation for science.

Conclusion

We have seen that a generally agreed on set of necessary and sufficient conditions for something to count as science has not been found. That is why the statement by Larry Laudan at the beginning of the chapter should serve as a warning to those who attempt to define science so that it is clearly isolated from other disciplines. This popular understanding perpetuates a false notion of what science is and how it works.

Does that mean that we can never recognize a case of science or nonscience when we see it? No, it doesn't. But one does not need a definition of science to do this. We will return to this point at the end of the chapter, but for now, suffice it to say that, *in a general way,* science is a discipline that *in some sense* and *usually* appeals to natural explanation, empirical tests, and so forth. Nevertheless, the italicized words must be kept in mind. There are borderline cases that are hard to classify.

Science and Philosophy

As we have already pointed out, the definition of science is largely a philosophical question. In this regard, if one surveys most of the

35. See Moreland, *Scaling the Secular City*, pp. 43–75.

works in philosophy of science, one will almost never find an attempt at an airtight definition of science.[36] Science itself interpenetrates with other fields of study in ways that are difficult to characterize in a hard-and-fast way. In particular, science bears a very complex relationship to philosophy. To this subject we now turn.

What Is Philosophy?

Here we jump from the frying pan into the fire. There is no generally accepted set of necessary and sufficient conditions for determining what philosophy is. Philosophers can, perhaps, take consolation in the fact that at least the definition of philosophy is itself a philosophical question. In fact, this observation is both a key to understanding philosophy and a reinforcement of the point that the definition of science is a philosophical issue, at least in part. For philosophy is, among other things, a second-order discipline that investigates other disciplines. It is possible to have a philosophy of x, where x refers to science, language, education, and so forth. For example, the question *What is education?* is a philosophical question. A scholar trained in education may attempt to answer it, but when he does, he dons the philosopher's cap. His answer will involve theories of truth, value, purpose in life, cognition, the nature of being human, the role education ought to assume in culture, and so on.

Since our focus is not directly on philosophy but on science, we do not need to construct an airtight definition of philosophy or even one that holds in most cases. For our purposes, we can content ourselves with giving illustrations of typically philosophical questions.[37] Philosophers are interested in questions about the nature of existence, truth,

---

36. Philip Kitcher's work *Abusing Science: The Case Against Creationism* (Cambridge, Mass: MIT Press, 1982), pp. 30–54, is a good example of this. He offers a generalization of science ("If a doctrine fails sufficiently abjectly as a science, then it fails to be a science. Where bad science becomes egregious enough, pseudoscience begins." p. 48), but wisely refrains from tightening the definition. The lesson to be learned is this: If scholarly experts in the field cannot define science so as to rule out religious considerations by definition, then the popular mythology of the separation of science and religion is hopelessly naive. Instead, what is needed is an examination of scientific and religious issues on a case-by-case basis. Such an examination reveals a complex interaction through history that gives the lie to the plethora of protests against creation science that rely on definitions of science and religion. For more on this, see David Lindberg and Ronald Numbers, eds., *God and Nature: Historical Essays in the Encounter Between Christianity and Science* (Berkeley: University of California Press, 1986).

37. Cf. McMullin, "Alternative Approaches to the Philosophy of Science," pp. 3–19.

mind, causation, space, time, rationality, value, perception, rational justification of theories, explanation, freedom, life, purpose, substance, properties, and so on. Clearly several items on this list interface with science (e.g., the nature of space and time, the nature of theories and explanation, and the nature of life). Let us look at different ways philosophy relates to science.[38]

Philosophy and Science

There are several ways of understanding the relationship between philosophy and science.[39] Rather than survey all of them, we will focus on three broad relationships between the two.

A FOUNDATION FOR SCIENCE

Philosophy operates underneath science in a foundational way. Philosophy is foundational to science in at least three ways.

First, attempts to put limits on science are philosophical in nature. As we have already seen, questions involved in defining science or stating necessary and sufficient conditions for demarcating science from nonscience are largely philosophical, as are questions about the nature of definition itself. Similarly, questions about integrating science and theology are largely philosophical and not directly scientific or theological, although the latter are obviously involved.

To see this, consider the following statements:[40]

(S1) Religious beliefs are reasonable only if science renders them so.

(S2) Religious beliefs are unreasonable if science renders them so.

(S3) Religious beliefs are reasonable only if arrived at by something closely akin to the scientific method.

38. For more on this, see Losee, *Historical Introduction*, pp. 1–4; Keith Campbell, *Metaphysics: An Introduction* (Encino, Calif.: Dickenson, 1976), pp. 1–22.

39. I am not considering two extreme ends of the continuum about the relationship between philosophy and science. At one end are views like W. V. O. Quine's, which see philosophy as essentially a descriptive, scientific enterprise. But this view fails to capture the normative, foundational nature of philosophical questions and borders on being self-refuting. See Carl Kordig, "Self-Reference and Philosophy," *American Philosophical Quarterly* 20 (April 1983): 207–16. At the other end are those who see science as focusing on mere appearances and philosophy as focusing on reality. More will be said about this in chapters 4 and 5.

40. I owe this insight to an unpublished paper by Keith J. Cooper entitled "Religion Within the Limits of Science Alone?" given at the Conference on Science and Christianity sponsored by the Institute for Advanced Christian Studies held at Madison, Wisconsin, in August 1986.

(R1) Scientific beliefs are reasonable only if religion renders them so.

(R2) Scientific beliefs are unreasonable if religion renders them so.

(R3) Scientific beliefs are reasonable only if arrived at by religiously appropriate methods.

These six principles attempt to place limits on science and religion. But none of them is an example of either science or religion directly limiting the other, for none is a statement *of* science or religion. Rather, all are philosophical statements *about* science and religion. Principles about science and religion are not at the same time results of or principles within science and religion. The six principles listed are philosophical attempts to limit science and religion.

Second, philosophy undergirds science by providing its presuppositions. Science (at least as most scientists and philosophers understand it) assumes that the universe is intelligible and not capricious, that the mind and senses inform us about reality, that mathematics and language can be applied to the world, that knowledge is possible, that there is a uniformity in nature that justifies inductive inferences from the past to the future and from examined cases of, say, electrons, to unexamined cases, and so forth. These and other presuppositions of science will be discussed in chapter 3, but all of them are philosophical in nature.

Third, philosophy undergirds science by focusing on what constitutes good theories, good cases of scientific explanation, confirmation, and so on. What makes a theory worthy of belief? How do theoretical values—simplicity, empirical accuracy, scope of application, clarity of concepts, and predictive success—figure into the evaluation of theories? What is a good explanation and how should science explain things? Should a scientist construct models, say, a billiard-ball model of an ideal gas, or should she merely subsume facts under general laws (all metal rods expand when heated, this is a metal rod, therefore this expanded when heated)? Should a scientist use only efficient causes (that by means of which something occurs), or should she also use functional or teleological explanations (that toward which or for the sake of which something occurs) that refer to future states (e.g., some intermediate states in chemical reactions obtain in order to diffuse an unbalanced electric charge as widely as possible)? How many positive instances of a law are needed to confirm it, or do positive instances offer no positive confirmation but merely show that the law has not been falsified yet? What is the cognitive status of scientific laws and theories? Do they really describe unseen entities, structures, and processes, or do they merely summarize and codify sensory experience and provide pragmatic guides in our attempt to control nature? Questions like these will be the focus of chapters 2, 4, and 5, and again, all are philosophical.

GOING BEYOND THE BOUNDS OF SCIENCE

Philosophy operates outside the domain of science. Sometimes a philosopher focuses her gaze on issues that lie outside science and have no immediate conceptual relevance to the scientific enterprise. For example, philosophers discuss questions about the nature of legal or natural rights. Are there universal positive human rights? How do positive rights differ from negative rights? What is a moral virtue, and how do we obtain moral knowledge? Is utilitarianism to be preferred to other normative ethical theories?

We need not illustrate further philosophical questions that lie outside science, but one point needs to be made. Our culture is so inundated with scientism—roughly, the view that only what science says is true or rational is, in fact, true or rational—and there has been such a pragmatic emphasis on science in education (with a concomitant devaluation of the humanities), that there is a widespread cultural myth that questions like those above are mere matters of private opinion. But that is just not so. Those questions and hosts of others are philosophical, lie outside science, and involve truth and rational justification. One need only examine examples of nontrivial discussions of these issues to satisfy himself of that claim. The same point could be made about theological debate.

INTERFACING DISCIPLINES

Philosophy operates within and interfaces with science.

Science does not consist merely of doing experiments. Scientists are involved in, among other things, developing an adequate and self-consistent set of concepts that aid us in understanding the world and solving various problems about the world. Thus, the practice of science involves philosophical reflection, and there are at least three ways that philosophy and science interact.

Philosophy helps to clarify internal conceptual problems in science. Internal conceptual problems arise when the concepts of a theory appear to be logically inconsistent, vague and unclear, or circularly defined.[41] These are all philosophical issues that require conceptual elucidation in the following ways.

41. See Larry Laudan, *Progress and Its Problems: Towards a Theory of Scientific Growth* (Berkeley: University of California Press, 1977), pp. 49–50.

1. The concepts of a theory are themselves problematic. An example would be the wave/particle nature of electromagnetic radiation and the wave nature of matter. These concepts appear to be self-contradictory or vague, and attempts have been made to clarify them or to show different ways to understand them. Another example concerns some conceptions of the mechanisms involved in evolutionary theory. Some scientists have held that evolution advances to promote the survival of the fittest. But when asked what the "fittest" were, the answer was that the "fittest" were those that survived. This was a problem of circularity within evolutionary theory, and attempts have been made to redefine the goal of evolution (e.g., the selection of those organisms that are reproductively favorable) and the notion of fitness to avoid circularity.

Another example is the discovery of the electron by J. J. Thomson (1856–1940) near the end of the nineteenth century.[42] At that time there was a debate between German and British scientists over the nature of electricity, the former favoring an aether wave view and the latter favoring a particle picture. Earlier in the century, Michael Faraday (1791–1867) had conducted various electrolysis experiments—experiments in which electric currents are passed through a water solution of decomposable compounds. He had shown that the amount of product liberated by such experiments is proportional to the amount of electricity introduced into solution and that the same amount of electricity liberates masses of products proportional to chemically equivalent weights.

But Faraday and others of his day had no clear way to understand these results because of their metaphysical conception of electricity as a continuous field to be treated as a wave. Thomson held to a different metaphysical conceptual scheme that viewed the basic units of electricity as particlelike atoms. This philosophical and metaphysical difference allowed Thomson to explain and further test Faraday's results. If one assumes that electricity is carried by tiny units called electrons, then more products will be liberated in electrolysis if more electrons are introduced into solution. Further, the fact that the amount of substance liberated was proportionate to the chemical equivalence instead of to the atomic weight indicates that substances with atoms of one positive charge require one "atom" of negative electricity, those with a positive charge of two require two "atoms" of negative electricity, and so on. It was not observational results alone that led to these discoveries, but also conceptual clarification about the nature of

42. See Harre, *Great Scientific Experiments*, pp. 156–65.

fields, particles, and so on that guided research, produced test implications, and allowed observations to be seen.

2. Philosophy elucidates the relationship between an operational definition in science, on the one hand, and a philosophical or ordinary-language definition, on the other. Scientists often define terms in operational language. That is, they define a term as referring only to a set of measurable operations or quantities that can be statistically tested, nothing more. For example, a psychologist may define depression as a mental state in which a person's responses fall between two scores on some standard test (like the MMPI). Or good sex may be defined operationally in terms of simultaneity of orgasm, intensity of certain physical parameters (increase in heart rate, skin tension, etc.), or the shape of a curve plotting the intensity of selected physiological parameters as a function of time.

Of course, depression and good sex mean different things in ordinary language. In fact, they may really be different from the operational factors used in the scientific definitions. Thus, an equivocation arises, because there are two different meanings to the terms (*depression* as it is used scientifically and *depression* as it is used in ordinary language). Scientists sometimes speak reductively as if all there is to some phenomenon is the operational aspects of the scientific definition. Philosophical clarity is needed to indicate the relationships between an operational definition and an ordinary-language or philosophical one. Is good sex really nothing more than having obtained certain measurable features in an experiment? If there is more to it, what is the relationship between the operational definition and the ordinary-language or philosophical one?

Consider another example. I once heard a psychologist argue that studies had shown that a mature adult was someone who had transcended traditional taboos, family relationships, and so forth. I pointed out that psychology could show nothing of the sort. The ordinary-language or philosophical notion of a mature adult is prescriptive, that is, normative. The psychological notion of a mature adult is merely descriptive and operationally defined as someone who falls within a certain range on a statistical curve describing a certain sample of people. The two definitions may be related in interesting ways, but they may also not be related, and in any case they certainly are not identical. Philosophical clarity is needed in attempts to identify or relate operational definitions and ordinary-language or philosophical definitions.

3. Philosophy helps to clarify claims associated with a scientific phenomenon. Sometimes philosophical claims are made in conjunction with some scientific theory. For example, some physicists working

in quantum theory hold that nature herself is statistical and indeterminate, not just our knowledge of nature. Others claim that the second law of thermodynamics applies not to the universe as a whole but only to specified regions of the universe that can be clearly defined. Again, scientists like Isaac Asimov and Paul Davies claim that the universe could have come into existence from nothing, and part of the support for their claim is the contention that matter can be produced from a state of zero energy where positive and negative energy are balanced.[43] Our purpose here is not to evaluate these claims, but merely to point out that when scientists make general statements like the ones just mentioned, even though these claims are associated with, and in some sense derived from, certain scientific theories, philosophical considerations are relevant in assessing them.

4. Philosophy assists in assessing categorial aspects of scientific claims. Sometimes scientific theories treat some phenomenon as an example of a certain category of thing. For example, heat used to be treated as an example of the category called substance—it was viewed as a physical substance or stuff, much like some kind of fluid, that flowed into or out of bodies that warm up or cool off. Later, heat was viewed as in the category of quality—a property of a fluid. Later still, heat was viewed as in the category of quantity—in some cases, the mean kinetic energy of the particles of a gas.

Roughly the same thing has happened to the concept of color. Colors used to be treated as qualities, but there is a tendency in science to identify color with wavelength and treat it as nothing more than a certain frequency of light.

Sometimes scientists assert that certain identities are true in nature, for example, that the stuff we drink, swim in, and cook with is really identical to H_2O. In this case, water is being treated as a *compositional stuff*. Philosophers examine the nature of identity as it relates to compositional stuff. For example, it seems to be a philosophical intuition that if something like water gets its nature by being composed of certain essential constituents (hydrogen and oxygen in a specific arrangement), then nothing will count as water without this composition, even if it looks, tastes, and feels like water, and even if we cook with it and swim in it.

Now consider an example of something we would call a *functional stuff*.[44] Suppose the only food in England was mutton stew and all the mutton stew in England was food. Now suppose people from England

43. Cf. Moreland, *Scaling the Secular City*, pp. 34–41.
44. See Bealer, "The Philosophical Limits of Scientific Essentialism" (see note 19), pp. 289–365.

traveled to the United States and were served hamburgers. Should they phone home and complain that they were offered no food in the U.S.? No, because food is not primarily a *compositional stuff*, but a *functional stuff*. Something is food if it is served at meals, sold in restaurants, ingested for nutrition, and so forth. The identity conditions for something being the same functional stuff as something else are different from those for its being the same compositional stuff.

These examples illustrate the philosophical role of categorial aspects of scientific claims. Is light really nothing but a wavelength, or is it merely that one can only have a certain shade of color if and only if a certain frequency of light is present? Perhaps light causes color but is not identical to it. These two views of color classify it differently (as a quantity or quality), they are empirically equivalent (no observation could judge between them), and they involve philosophical considerations. Similarly, claims that water is H_2O involve philosophical assumptions about appropriate ways of viewing identity for compositional entities. Is a baboon heart that has been transplanted into a human a human heart or still a baboon heart? That depends on whether the biological notion of a heart is compositional or functional. It is still composed of the same stuff that baboon hearts are made of, but it is now functioning in a human body. Issues involved in claims like these involve philosophical clarification and analysis because they are examples of categorial classification.[45]

5. Philosophical concepts provide pictures to guide research and to derive scientific theories and test implications. Often a certain metaphysical concept of reality can provide a picture that guides scientific research and allows the scientist to derive theories or testable implications from the concept. For example, the philosopher Gottfried Leibniz (1646–1716) was able to derive Snell's law (a ray of light traveling obliquely, i.e., neither parallel nor perpendicular, from one optical medium into another is refracted at the surface in such a way that the ratio of the sines of the angles of incidence and of refraction is a constant for any media pair—$n_1 \sin \Theta_1 = n_2 \sin \Theta_2$) from his metaphysical principle that nature always selects the easiest, most direct course of action given a set of alternatives. Leibniz derived this metaphysical principle, in turn, from his theological conviction that God created the world such that a maximum of simplicity and perfection should be realized.

---

45. For an example of how the notion of identity relates to a comparison of causes and effects in Michael Faraday's identification of various forms of electricity (produced from such diverse effects as friction, heat, magnetism, chemicals, and animals), see Harre, *Great Scientific Experiments*, pp. 176–84.

Linnaeus and other early taxonomists who worked on classifying organisms were guided by a combination of Christian theology and Aristotelian philosophy that produced the conviction that living things had fairly fixed essences that would permit them to be organized into a hierarchy of classes. These and many other examples show that philosophical and theological conceptions have guided scientific research and provided grounds for deriving theories and testable propositions.

Philosophy often examines scientific entities in a more general way than does science. Often scientists and philosophers approach the same entity from a complementary point of view, philosophy asking more general questions than science. These two approaches to describing an entity need not conflict unless one of the disciplines takes its approach as the whole truth or unless the disciplines make competing claims at the same level of description.

For example, scientists often describe electrons in certain ways (as having negative charge, a certain rest mass, and so on). When they talk about electrons, they treat them as all of exactly the same nature. All electrons have all and only the same properties and they differ from one another only numerically. But if two electrons, A and B, share literally all of the same properties, what makes them two distinct electrons instead of just one? This is called the problem of individuation, and it is properly a philosophical problem. Different philosophers solve the problem differently—A and B are two entities because they are located in different places; each has one property the other does not have (being identical to A, and being identical to B); each has, in addition to its properties, another entity called particularity, or a bare particular, or a pure, numerically primitive thisness. In this case, philosophical theories of individuation try to solve a more general issue regarding electrons than do scientific approaches. And some solutions—for example, the bare particular theory (which I favor)—involve postulating an entity in electrons that is metaphysical in such a way as to be beyond the bounds of scientific investigation. In this case, conflict arises when science claims to say all that needs to be said about electrons.

As an example of a conflict that arises when scientists overstep their bounds, consider the following statement by biologist E. Mayr:

> The concepts of unchanging essences and of complete discontinuities between every eidos (type) and all others make genuine evolutionary thinking well-nigh impossible. I agree with those who claim that the essentialist philosophies of Plato and Aristotle are incompatible with evolutionary thinking. . . . The assumptions of population thinking (evo-

lution) are diametrically opposed to those of the typologists.... The ultimate conclusions of population thinkers (evolutionists) and of the typologist are precisely the opposite. For the typologist, the type (eidos) is real and the variation an illusion, while for the populationists (evolutionists) the type (average) is an abstraction and only the variation is real. No two ways of looking at nature could be more different.[46]

Apart from some confusion about certain philosophical issues (abstractions are not unreal for most philosophers and Aristotle did not think variations were illusions), Mayr is confusing scientific ideas and philosophical ones and, thus, overstating the case. The philosophical notion of an essence (e.g., humanness, horseness, animality) is postulated to explain certain philosophical issues—finding a ground for the unity of classes of things (What unites the class of all and only humans?), grounding the unity of the parts and properties of a specific entity (What makes the hands and color of Jones a unity that includes his height but excludes his pen?), grounding a thing's sameness through change, and so on.

Now, the philosophical notion of an essence and the scientific description of an organism's taxonomic classification may interrelate. And the existence in organisms of philosophical essences may tend to count against evolutionary theory. But this is not clearly the case, since the two concepts of essence may be complementary aspects of an organism postulated to explain different problems. Whatever the case, Mayr's claim is philosophical, not scientific, and as such it illustrates both the importance of philosophical clarity about scientific definitions and terms and the fact that philosophy studies some of the same entities as does science albeit from a broader perspective that may be complementary in some cases and competitive in others.

Philosophy (with other disciplines) provides external conceptual problems for scientific theories.   Larry Laudan has argued that the history of science illustrates that conceptual difficulties from philosophy, theology, logic and mathematics, political theory, and general world-view considerations provide negative evidence for the rational acceptance of scientific theories in the form of external conceptual problems.[47] "Thus," he says, "contrary to common belief, it can be rational to raise philosophical and religious objections against a particular theory or research tradition, if the latter runs counter to a well-established part

46. Cited in Michael Denton, *Evolution: A Theory in Crisis* (London: Burnett Books, 1985), pp. 98–99.
47. See Laudan, *Progress and Its Problems*, pp. 45–69, especially pp. 50–54.

of our general *Weltbild*—even if that *Weltbild* is not 'scientific' (in the usual sense of the word)."[48]

This observation makes sense. If science really is not an isolated discipline tucked away in an airtight compartment, if one has good arguments or reasons for holding to some proposition, and if a scientific theory conflicts with that proposition and is not merely complementary to it, then the proposition itself provides some evidence against the scientific theory. This is so even when the proposition in question is theological, philosophical, or related to some other discipline outside science. The real issue is not what kind of proposition it is, but how strong the evidence is for it.

Anyone wanting an integrated world view will see that nonscientific problems, if they are rationally supportable, will count against conflicting (though not complementary) scientific claims; thus, such external problems should be used in assessing those scientific claims. Put another way, by its very nature, science interacts and sometimes conflicts with other rational fields of study. Thus, it is not inappropriate to the nature of science or to the way science has been practiced through history to raise theological or philosophical considerations as part of the assessment of a scientific theory, if those considerations are rationally justifiable on their own grounds.

An external conceptual problem arises for a scientific theory T when T conflicts with some doctrine of another theory T', when T' and its doctrines are rationally well founded, regardless of what discipline T' is associated with. T may be logically inconsistent with T' or one of its doctrines, or the two may conflict in a lesser way by being jointly implausible (though still logically compossible), that is, merely compatible.

External conceptual problems arise in at least three ways:

1. Conceptual problems may arise from a discipline outside of science but still be legitimately a part of science.

For instance, Ptolemy's system of astronomy conflicted with general contemporary metaphysical views (which were rational beliefs at the time) that celestial motion should be perfect, that is, perfectly circular at constant speed. Leibniz and Berkeley raised conceptual problems with the meanings and intelligibility of absolute space, absolute motion, attractive and cohesive forces, and so on. As mentioned earlier, J. L. Mackie has raised a serious philosophical objection to the special relativity notion that there is no such thing as an absolute reference frame for absolute rest and absolute motion.

The Jesuit scientist Ruggiero Boscovich (1711–1787) raised serious

---

48. Ibid., p. 124.

difficulties with the Newtonian conception of matter that were never adequately answered. Boscovich showed that if atoms were taken to be solid, impenetrable bodies (i.e., of infinite hardness) with no compression on impact, then the following problems arise: there is no time interval, $\triangle$ t, *during* impact between two atoms; changes of momentum *at* impact are spontaneous; and a change in momentum of an atom requires it to be simultaneously at motion and at rest!

Philosophers also raised problems with the concept of action at a distance. Newton postulated two kinds of forces: the force of impact, when two bodies contact each other, and gravitational force, which can operate at a distance without two bodies being in contact. Philosophers such as René Descartes (1596–1650) pointed out that it would be a simpler theory if the latter force could be reduced to some type of force involving contact and that there was no clear conception of causation that did not involve such contact. This conceptual problem did not go away, and it led to further study of gravity. Today many scientists deny action at a distance and see the gravitational force between spatially separated bodies as due to an intermediate field or transmitted by an exchange of particles (called gravitons) between the bodies.

Finally, some philosophers, myself among them, have raised philosophical difficulties against the existence and traversability of an actually infinite number of events.[49] These philosophical arguments count against any cosmological model that postulates an infinite past, for instance, the oscillating universe theory.

2. Conceptual problems may arise if science is taken to be the whole story on some phenomenon and such a posture undercuts the necessary preconditions for science itself to exist.

A number of philosophers, like Keith Lehrer, have argued that certain varieties of physicalism regarding the mind/body problem that deny the existence of mind are self-refuting.[50] Physicalism, if taken to be the only word about the nature of a human mind, as opposed to a limited approach to the brain, undercuts the necessary preconditions for rationality itself.

Similar arguments have been raised against some sociologists who have attempted to reduce intellectual history—the study of the rational aspects of systems of ideas and how they change through history by their own inner logic (e.g., how John Locke's [1632–1704] form of empiricism naturally gave rise to Berkeley's form and the latter to

49. See Moreland, *Scaling the Secular City*, pp. 18–33.
50. Keith Lehrer, *Knowledge* (Oxford: Clarendon, 1974), pp. 236–49. See also Moreland, *Scaling the Secular City*, pp. 90–96.

David Hume's [1711–1776] form)—to the sociology of knowledge. Sociology of knowledge studies nonrational factors that have influenced the change of ideas through history and in different cultures—the economics or the form of government at the time. However, if sociology of knowledge is the whole story, then the rational/nonrational distinction collapses and so does the rationality of the sociologist's own theory.

Similarly, some have raised the same arguments against atheistic forms of the theory of evolution. Charles Darwin (1809–1882) himself mused about why one ought to trust the deliverances of the mind if it were a mere product of a blind process of natural selection and survival.[51] The details of these arguments are not important here. What is important is that such arguments provide external conceptual problems for certain scientific theories if they are taken to be the whole story of some phenomenon.

3. Conceptual problems may arise when some scientific theory, while strictly consistent with some area of knowledge outside of science in terms of logic, still tends to count against it, even when falsifying that area of knowledge would not necessarily undercut science itself.

An example of this kind of problem can be seen in theories about evolutionary ethics. If one accepts atheistic versions of evolutionary ethics, then this provides strong evidence that no moral properties or virtues constitute the natures of acts and things (e.g., persons), and that there are no objective, absolute moral principles. If one sees the entire cosmos as coming from a blind explosion called the Big Bang and life arising through a blind, natural process of mutation and struggle for reproductive advantage, it is hard to see what could give man moral worth. If morality arose in the context of such a picture, and if morality does not reflect an objectively existing moral order but merely mirrors social and psychological rules grounded in the instinct to survive, then this provides strong evidence against our common-sense intuitions about the nature, form, and objectivity of morality.

Of course, many thinkers have argued that how morality arose is a different issue from what morality is and that moral language has

51. See Darwin's letter to W. Graham dated July 11, 1861, cited in Stanley L. Jaki, *Angels, Apes, and Men* (La Salle, Ill.: Sherwood Sugden, 1983), p. 57. For more on how Darwin wrestled with this problem, see *Metaphysics, Materialism, and the Evolution of Mind: Early Writings of Charles Darwin*, transcribed and annotated by Paul H. Barrett (Chicago: University of Chicago Press, 1974). A good, brief secondary source on this issue is Jaki, *Angels, Apes, and Men*, pp. 51–72. It is interesting to read some current discussions of how to justify science given the problem of mind and rationality within the framework of a naturalistic interpretation of evolution. See Bas C. van Fraassen, *The Scientific Image* (Oxford: Oxford University Press, 1980), pp. 39–40.

a different logic and nature (e.g., it is prescriptive) from scientific language (e.g., it is descriptive), and so on. There is a point to this rejoinder. But our common-sense intuitions about morality and their more sophisticated philosophical counterparts in natural law and deontological theories still seem incompatible with atheistic evolution. The latter does tend to disconfirm these intuitions and theories by providing a background theory of the world within which the existence and nature of objective morality appear puzzling, especially when atheistic evolution is compared to various theistic accounts of the origin and development of the universe.

So the existence and nature of morality as it is normally conceived present external conceptual problems to atheistic evolution, though perhaps not to theistic evolution. One can abandon common-sense morality, one can embrace theistic versions of evolution or other forms of creationist theories, or one can hold common-sense morality and atheistic evolution together. But the last approach occasions an uneasy tension because the two are merely compatible, not mutually reinforcing but in tension with each other, and the theistic alternatives cited provide a better, more justificative background theory of the world within which common-sense morality is not merely possible, but likely and reasonable. Thus, common-sense morality, evolutionary ethics, and atheistic versions of evolutionary theory clash. Such a clash illustrates this third kind of external conceptual problem.

Before summarizing our discussions in this chapter, one caveat about conceptual problems is in order. Conceptual problems are not the only problems relevant to scientific theories; neither are they in themselves sufficient to warrant a suspension of judgment or an abandonment of some scientific theory. But they are a legitimate part of scientific practice and theory assessment. The role and weight given to a conceptual issue should be determined on a case-by-case basis. But to admit this much is to admit that theological and philosophical issues can be a part of the very nature of science. Judge Overton and those of his persuasion are sadly misinformed.

Summary and Conclusion

It is time to gather our results and state some conclusions.

1. There is no clear-cut definition of science. Neither are there any generally accepted necessary and sufficient conditions for drawing a line of demarcation between science and nonscience. It is foolish to say, with popular opinion, that science by definition rules out theo-

logical or philosophical concepts. One can recognize clear examples of science without a definition (acid/base titration measurements), especially if one does not ask too many questions about the example at hand (in which case philosophical issues will be inseparably involved), and clear cases of nonscience (palm reading and astrology). But these are at opposite ends of a continuum with fuzzy boundaries and several borderline cases. In particular, one cannot rule out creation science by definition. Such a position is question begging and codifies a historically naive, philosophically arcane view of science. More on this in chapter 6.

2. The question of defining or stating necessary and sufficient conditions for science is in large measure philosophical. This is true even if the person defining science is a scientist acting as a philosopher. The view that scientists are uniquely qualified to define science is due in large measure to an educational system that emphasizes pragmatic considerations, stresses analysis instead of synthesis, and deemphasizes the humanities. Scientists themselves are most often the last people to ask to define science. In earning my B.S. in chemistry, I observed that serious study in the history and philosophy of science was singularly absent from most science curricula. The scientist is trained in first-order practices of studying amoebas, quarks, and the like. He is not trained in the second-order practice of studying science as a discipline.

3. Theological concerns are important to science at several levels. One of the main ways theology interfaces with science is by providing rational conceptual problems for scientific theories. This often is achieved via philosophy; for example, the theological concept of God as Creator is studied as a philosophical concept (an agent who acts in various ways that can be specified and from which conceptual and empirical implications for science can be drawn). In general, the integration of science and theology is a philosophical question.

We have seen that science cannot be separated from other disciplines by stating necessary and sufficient conditions for science. Science is not an isolated, airtight field of study. But perhaps this conclusion is a bit premature. So far we have said very little, apart from our discussion of conceptual problems, about scientific methodology. Several thinkers believe that somehow science differs from nonscience by its use of the so-called scientific method. As Ernest Nagel has said, "... the conclusions of science, unlike common-sense beliefs, are the products of scientific method. ... The practice of scientific method is the persistent critique of arguments, in the light of tried canons for judging the reliability of the procedures by which evidential data are obtained, and for assessing the probative force of the evidence on which conclusions are based."[52] Issues involved in clarifying the so-called scientific method will occupy our attention in the next chapter.

52. Ernest Nagel, *The Structure of Science* (Indianapolis: Hackett, 1979), pp. 12–13.

Scientific Methodology

> ... the much sought-after "scientific-method" may be a will-o'-the-wisp.
>
> —Larry Laudan, *Science and Values*

> At the outset it should be stated that there is no "*scientific method,*" no formula with five easy steps guaranteed to lead to discoveries. There are many methods, used at different stages of inquiry, in widely varying circumstances.
>
> —Ian G. Barbour, *Issues in Science and Religion*

There is a fairly widespread belief that there is something called *the* scientific method that can be characterized in a fairly clear, unequivocal manner and that separates science from other fields. For example, the following statement occurs at the beginning of a widely used high school biology text: "Scientists use the scientific method in attempting to explain nature. The *scientific method* is a means of gathering information and testing ideas. . . . The scientific method separates science from other fields of study."[1] This stereotype is, unfortunately, both false and widely believed to be true.

In this chapter we will investigate a number of issues that arise in the context of characterizing scientific methodology, and we will discover two things: First, there is no such thing as *the* scientific method, but rather there is a cluster of practices and issues that are used in a variety of contexts and can be loosely called scientific methodologies. Second, various aspects of scientific methodologies arise in the practice of disciplines outside science. These two discoveries will reinforce our conclusion, in chapter 1, that science is not isolated from other disciplines.

These conclusions should not surprise us. Belief that the scientific

---

1. Peter Alexander, et al. *Biology: Teacher's Edition* (Morristown, N.J.: Silver Burdett, 1986), p. 4.

method is a single, clear-cut procedure used by science alone is naive for at least two reasons.

First, as we have already seen, science is not an airtight compartment isolated from philosophy, theology, and other disciplines. It interacts with other fields of study in complicated ways. Only someone out of touch with the history of science could believe otherwise. When Copernicus offered his heliocentric theory as an alternative to Ptolemy's elaborate mathematical scheme of heavenly motion, part of the case against Ptolemy was his violation of the Platonic ideal of circular motion and the Aristotelean picture of spheres within spheres.[2] The point is not that these considerations were determinative, but that they were rational considerations for thinkers in the 1500s. Issues of this kind have always played an essential role in science.

Second, belief in a single method unique to science is naive because it arises from a superficial understanding of human action. The formation, use, and justification of scientific ideas are particular examples of human actions. As such, they partake of features of human actions in general.

Consider two people, Walt and Marty, who individually visit their uncle Louis on consecutive days. If you were hiding in the closet watching, and if you didn't ask too many questions or think very deeply about what was going on, you might conclude that each did the same thing: each visited Louis, spent time with him, and cheered him up. However, closer inspection might have revealed that, appearances to the contrary, they did different things. Walt, motivated by hate for his father and a desire to influence Louis to include him in his will, and knowing that Louis thought poorly of him, tried to secure a place in Louis's will by bragging about his business sense, his responsible character, and so on. But Marty, motivated by compassion and knowing that Louis loved rhubarb pie, showed love by satisfying Louis's physical appetites. If your approach to Walt and Marty stayed at the surface and you didn't ask too many questions, you might get the mistaken impression that each did the same thing. After all, their actions looked the same at one level. But in reality, their actions differed at the levels of motive, intent, and means.

Similarly, if you don't ask too many questions about two people in a chemistry laboratory, then you can get the impression that they are doing the same thing. Both put an acid in a beaker and read the pointer after thirty seconds. But if you carefully examine what each

2. A helpful treatment of the rational factors of the Copernican debate that is sensitive to the historical context is Colin A. Russell, *Cross-Currents: Interactions Between Science and Faith* (Grand Rapids: Eerdmans, 1985), pp. 22–53.

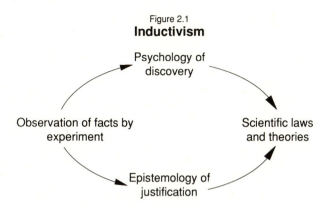

Figure 2.1
Inductivism

sees herself as doing, then you could see that one is trying to falsify a hypothesis while the other is trying to verify a hypothesis. One is trying to "save the phenomena" while the other is trying to describe the deep structure of allegedly real molecules.

Differences of this type, and a host of others, must be considered if we are to examine scientific methodology any way but superficially. When such issues are brought to the surface, a family of scientific methodologies arises. In what follows, we will look at one way of viewing "the scientific method" called inductivism; analyze different facets of a more appropriate picture of scientific methodology; offer an illustration of scientific methodology used outside of science.

Inductivism

Inductivism is a view of the scientific method wherein scientists are seen as progressively piling up more and more facts by observations, generalizing them by enumerative induction into laws, combining these generalizations into broader and broader generalizations by piling up more facts, and, finally, arriving at various levels of scientific laws whose contents are nothing but the facts. Inductivism pictures the scientific method as in figure 2.1.[3]

The psychology of discovery refers to the temporal process an individual scientist, or a community of scientists, uses in step-by-step

3. Several different methodologies go by the name of induction, but essentially induction is a form of inference wherein the truth of the premises does not guarantee the conclusion the premises support but only makes it probable to various degrees of adequacy. Enumerative induction is a type of inductive inference that starts by listing specific examples of a phenomenon (cat_1 is black, cat_2 is black . . ., cat_n is black) and ends by generalizing to a conclusion (all cats are black or x% of cats are black).

fashion in forming laws or theories. The epistemology of justification refers to the normative, logical structure by which a scientist, or a community of scientists, justifies scientific laws or theories. Inductivism is a thesis about both the psychology of discovery and the epistemology of justification.

As a thesis about the psychology of discovery, inductivism implies that the scientific method starts with observations and experiments and proceeds to the formation of laws and theories by enumerative induction, that is, by inductively deriving generalizations from past observations. As a thesis about the epistemology of justification, the inductivist position holds that a scientific law or theory is justified only if the evidence in its favor fits the inductive schema.[4]

According to advocates of inductivism, scientists form and test laws as follows: 1) They start with observations and experiments and record all facts without a prior, biased guess as to their relative importance. 2) They analyze and classify the facts. 3) They inductively derive generalizations from them. 4) They test these generalizations further by observations and experiments and form higher-order generalizations. Scientific knowledge is a conjunction of well-attested facts that grows by the addition of new facts that usually leave previous facts unaltered.

In sum, inductivism implies that the scientific method involves the following: 1) a psychology of discovery that starts with experimental observations without prior guesswork that might bias those observations; 2) an epistemology of justification wherein a) laws and theories are justified only if they are formed by an inductive schema, and b) our belief in the degree of plausibility of a law increases in proportion to the number of observed positive instances of the phemonemon described in it; 3) a view of scientific laws and theories as inductive generalizations that describe facts (or sensory experiences) and their regular relations; and 4) a view of experimentation wherein experiments provide the raw material, or data, for erecting the scientific edifice of laws and theories.

John Stuart Mill (1806–1873) is often associated with inductivism. Mill offered five canons, or principles, regulating the discovery and justification of experimental inquiry.[5] The two most important of these

---

4. For more on inductivism, see Carl G. Hempel, *Philosophy of Natural Science* (Englewood Cliffs, N.J.: Prentice-Hall, 1966), pp. 10–18; Rom Harre, *Philosophies of Science: An Introductory Survey* (Oxford: Oxford University Press, 1972), pp. 35–48, and *Great Scientific Experiments: Twenty Experiments that Changed Our View of the World* (Oxford: Oxford University Press, 1983), pp. 5–10.

5. For a good overview of Mill's inductivism, see John Losee, *A Historical Introduction to the Philosophy of Science*, 2d ed. (Oxford: Oxford University Press, 1980), pp. 148–58.

TABLE 2.1 **The Methods of Agreement and Difference**

The Method of Agreement

Instance	Antecedent Circumstances	Phenomena
1	ABEF	abe
2	ACD	acd
3	ABCE	afg

The Method of Difference

Instance	Antecedent Circumstances	Phenomena
1	ABC	a
2	BC	—

canons were the method of agreement and the method of difference, illustrated in table 2.1.

The method of agreement attempts to state sufficient conditions for some phenomenon. In table 2.1, A is the only factor present in the three instances and a is present as well. B–F can be eliminated and, thus, are not connected to a. The method of difference attempts to state necessary conditions for some phenomenon. In table 2.1, where A fails to obtain, a fails to obtain as well. And whatever cannot be eliminated without loss of an associated phenomenon must be connected with that phenomenon.

An alleged example of inductivism is the work of Gregor Mendel (1822–1884), the scientist-priest who studied the laws of inheritance by focusing on peas.[6] Mendel studied different traits of pea plants that appeared in two different forms: seed texture (smooth/wrinkled), seed color (yellow/green), and stem length (long/short). He and his assistant grew pea plants, crossbred them, and counted the number having certain characteristics in each generation. Some of Mendel's results appear in table 2.2. On the basis of these and other observations, Mendel allegedly derived, by pure induction, the following scientific law: In the second generation the ratio of dominant to recessive characters of the same allele (a gene that governs an expressed trait or phenotype that comes in alternative forms or genotypes) is 3:1.

Although inductivism continues to persist in the popular conception of science and even in the minds of many scientists, very few, if any, philosophers of science accept it. It fails to grasp the variegated texture of science, and the objections against it are severe.

First, one cannot merely start with observations without some guid-

6. Cf. Harre, *Philosophies of Science*, pp. 36–37; Karen Arms and Pamela Camp, *Biology*, 3d ed. (Philadelphia: Saunders College, 1987), pp. 281–87.

TABLE 2.2 **Some of Mendel's Results**

Parental Characters	F_1*	F_2*	F_2 Ratio
Smooth x wrinkled seeds	all smooth	5474s:1850w	2.96:1
Yellow x green seeds	all yellow	6022y:2001g	3.01:1
Long x short stems	all long	7871:277s	2.84:1

*F_1 and F_2 are the first and second generations, respectively.

ing hypothesis or background assumptions to guide in deciding what is and is not relevant to observe. Pure, presuppositionless observations are a fable in science, as illustrated by a story told by Karl Popper:

> Once upon a time there lived a man who wished to give his whole life to science. Noble creature that he was, this man sat down with pencil in hand and recorded in a notebook everything he could observe. He included everything, from the weather, the racing results, and the levels of cosmic ray bombardment, to the stock market reports and the appearance of all the planets. He also did not neglect to record the attendance of all students at weekly lectures. Our dedicated observer continued this job every day for the rest of his life. He had compiled, by the time of his death, the most comprehensive record of nature ever made in the history of mankind. When he died, he was so certain that his life had been well spent for the cause of science, he donated his notebooks (there were hundreds of them as you might suppose) to the American Academy of Science. But did the Academy bother to thank him for his donation and efforts? Did the Academy even accept his donation? No! They refused even to open his notebooks, because they knew without looking that they would contain only a jumble of disorderly and useless items.[7]

Scientists do not start with observations. They usually start with a problem to be solved and a set of assumptions and hypotheses about what is and is not relevant to observe.

Second, the same point can be made about classifying and arranging observations. Without some framework about what is going on, there is no way to decide what factors should serve as the basis on which to classify particular facts.

Third, scientific laws are not formed, accumulated, or justified by the progressive addition of brute, uninterpreted observational data. Rather, these processes involve a mixture of observation and theory in several different ways. Sometimes a shift in theory can turn seeming facts into falsehoods. As Harre has pointed out:

7. Cited in V. James Mannoia, *What Is Science? An Introduction to the Structure and Methodology of Science* (Lanham, Md.: University Press of America, 1980), pp. 14–15.

For instance, consider the history of the determination of the atomic weights. What *were* the facts? Under the influence of Prout's hypothesis some chemists considered that the discrepancies between integral values for the atomic weights of the elements were errors, since Prout had maintained that all elemental atoms were combinations of whole numbers of complete hydrogen atoms, and hence their atomic weights had to be integral numbers by comparison with hydrogen. Those who did not accept or had abandoned Prout's hypothesis were inclined rather to suppose that the non-integral weights were the facts, that is a genuine measure of a natural phenomenon. What the facts were depended in part upon whether one held or did not hold a particular theory.[8]

Fourth, inductivism pictures the formation and justification of scientific laws after the pattern of plotting points on a graph and curve fitting. Consider figure 2.2. Suppose we are trying to formulate a law, like Charles' law, which relates the temperature and volume of a gas. Four experimental observations have been taken and plotted on our curve. The difficulty is that there is a set of potentially infinite and empirically equivalent curves consistent with these points. Three are plotted in the middle chart. Which one is correct? The issue cannot be settled by plotting a fifth observation at e. This may rule out some curves but a potentially infinite number still survive. Further, we know that the pressure of the gas must be held constant for our points to be good in figuring out the temperature/volume relationship, and we know this background assumption by means of our theory of what gases are and how they behave.

Someone could respond by saying that we should draw the simplest curve as the generalized law of temperature and volume. But several problems are involved with this suggestion. 1) The idea that nature will, in general, behave in a simple way is not derived from observation but is presupposed. Such a presupposition is not properly a part of scientific methodology, according to inductivism. In fact, the simplicity assumption was formulated in the context of Christian theism, where simplicity in nature was seen as a part of the perfection of God. 2) While simplicity is often a good criterion to follow, sometimes a scientific law becomes more complex over time. The history of the study of gases shows a shift from the ideal gas law ($PV = nRT$) to the Van Der Waals' equation ($[P + a/V^2][V-b] = nRT$). 3) There are several ways to interpret simplicity (e.g., lowest power of ten, lowest number of independent variables, smoothest graph on polar coordinates versus Cartesian coordinates). Thus, simplicity is itself not simple, and even

8. Harre, *Philosophies of Science*, p. 43.

Figure 2.2*
Examples of Inductivism

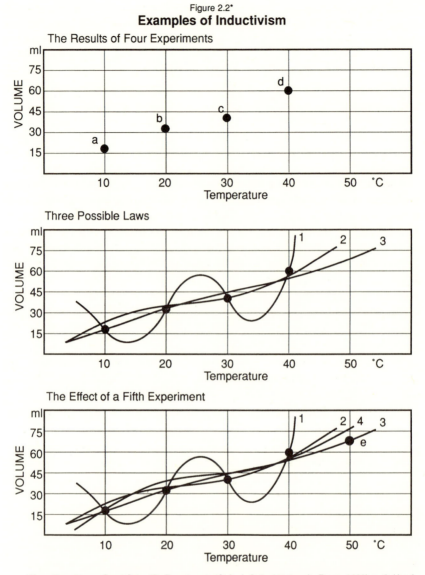

The Results of Four Experiments

Three Possible Laws

The Effect of a Fifth Experiment

*From Rom Harre, *Great Scientific Experiments* (Oxford: Oxford University Press, 1983), p. 8. Used by permission.

if it were, simplicity considerations are outside of inductivism as that doctrine is usually represented.

Fifth, there is a difficulty for inductivism known as the raven paradox.[9] Suppose we are trying to confirm the law L_1, "All ravens are black," by appealing to positive instances of that law, namely, the discovery of a number of black ravens. Now, "All ravens are black" is logically equivalent to its contrapositive, L_2, "All nonblack things are nonravens." If an observation confirms L_1 it also confirms L_2, and vice versa. But that means that the discovery of a nonblack nonraven (e.g., a clean baseball) tends to confirm L_1. If this is accurate, then as one philosopher puts it, one could study birds without ever getting his feet wet, for the discovery of white objects in one's room confirms a generalization about ravens. The difference between discovering a white baseball and a black raven is hard to capture on the inductivist model.

Several other problems have been raised against inductivism: the problem of induction itself (what justifies the inference from *all observed As are B* to *all As are B*), the difficulty of deciding between accidental and lawlike generalizations, a decision that can be made against the backdrop of a causal theory that is not a mere summary of observational facts (e.g., *plants grow from the sun's warmth* versus *plants grow from the sun's light*, the former an accidental generalization and the latter a causal law believed, in part, on the basis of theories about how photosynthesis works), and the fact that science does not try merely to describe phenomena by generalizations but also to explain them with theories about underlying mechanisms.

Inductivism is one view about what constitutes the scientific method, and it represents an indefensible misunderstanding of science.[10] In order to probe more deeply into scientific methodology, let us turn to a more eclectic model.

An Eclectic Model of Scientific Methodology

Figure 2.3 illustrates some of the complex interrelationships of elements in the scientific process. A close look at each element and its role will help us understand the process better.

9. See Carl G. Hempel, *Aspects of Scientific Explanation* (New York: Macmillan, 1965), pp. 14–20; Peter Achinstein, *The Nature of Explanation* (Oxford: Oxford University Press, 1983), pp. 367–74.

10. See Keeton, *Biological Science*, 3d ed. (New York: W. W. Norton, 1980), pp. 1–4. Keeton bemoans the fact that there is a surprising lack of understanding of what science is, but his own treatment of scientific methodology is filled with errors (e.g., "A theory is a hypothesis that has been repeatedly and extensively tested and always found to be true") and his view of science, if not identical to inductivism, is perilously close to it.

Figure 2.3
An Eclectic Model of Scientific Method

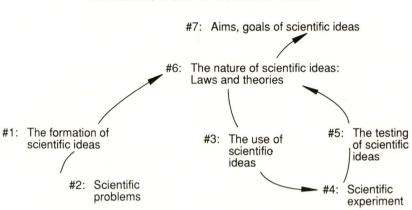

#7: Aims, goals of scientific ideas

#6: The nature of scientific ideas:
     Laws and theories

#1: The formation of   #3: The use of   #5: The testing
     scientific ideas   scientific   of scientific
                                    ideas   ideas

     #2: Scientific   #4: Scientific
          problems   experiment

#1: The Formation of Scientific Ideas

Let us begin our focus on a more eclectic understanding of scientific methodology by examining the psychological and, since science is practiced by communities, sociological process by which scientific ideas (i.e., laws, theories) are formed.

It is generally agreed that there is no formalized method, no step-by-step algorithm by which scientists form their ideas. Sometimes scientists discover things by accident.[11] On other occasions they generate their ideas in more bizarre ways. It is well known, for instance, that F. A. Kekule (1829–1896) came up with the hexagon formula for the benzene ring by having a trancelike vision of a snake attempting to chase its own tail and, thus, curving into such a ring.

More frequently, scientists generate their ideas by a creative process of educated guesswork, known as adduction. In adduction, a scientist does not form a theory by simple enumerative induction from observations to a generalization. Rather, by an act of creative imagination he adduces a conceptual web or gestalt, a conceptual pattern that

11. See the account of the Davisson-Germer electron scattering experiments in Arthur Beiser, *Concepts of Modern Physics*, 4th ed. (New York: McGraw-Hill, 1987), pp. 98–102. Davisson and Germer were studying the scattering of electrons off of a plate of nickel. At first their results were in keeping with classical predictions that electrons are merely particles (the scattering of electrons off of the random nickel surface was itself random with no significant dependence of intensity upon scattering angle or incident electron energy). But due to an accident in the experiment, the nickel target was reheated at high temperatures and the new defraction pattern exhibited maxima and minima in keeping with the De Broglie wave hypothesis.

models the phenomenon in question. For example, if light is shown onto the surface of a metal, electrons are emitted. If the frequency of the incident light is lowered, then a cut-off frequency is reached at which light of a low intensity (i.e., a low number of photons striking the surface per unit of time) will still liberate electrons, but below which even light of high intensity will not liberate electrons. This phenomenon is known as the photoelectric effect. Scientists had difficulty explaining the photoelectric effect within a gestalt that viewed light as a wave. But Albert Einstein (1879–1955) recommended a gestalt switch. That is, he viewed light as a series of particles, called photons, that carry bundles of energy proportional to the frequency of the light. An electron needs a minimum amount of energy to be liberated, and if an incident photon collides with it and has that energy, it will dislodge the electron. Einstein adduced the particle gestalt for light, which enabled scientists to explain the photoelectric effect.

When scientists adduce a scientific idea, they often do so as skilled craftsmen who have practical, tacit know-how about a domain of study. Much like a mechanic who just knows where to look for an engine breakdown, a scientist often has a feel, a sense for how things are in an area of investigation. Such a sense helps him to adduce theories to explain phenomena. The same sort of tacit knowledge is used by those outside science, for example, exegetes of Scripture who have a feel for Pauline literature, or spiritual mentors who, as Richard Foster puts it, have a feel for how one should read events in order to see or adduce God's working in and through them.

Sometimes scientists do not adduce an idea from tacit knowledge of a domain of study. Frequently in the history of science, they have derived their conceptual ideas from the metaphysical aspects of philosophical or theological theories. Philosophical or theological frameworks can motivate scientific investigation, guide research by suggesting lines of testing, provide conceptual problems to be solved, enable scientists to see data they could not see before, and determine, in part, what counts as veridical data (see the Prout example). Philosophical or theological theories do these things by helping to provide scientists with an *iconic model*, a picture of some supposedly real entity or process that behaves in ways dictated by the model. For example, a toy bear is an iconic model of a real bear, and a child can learn features of real bears by focusing on different aspects of the toy.

Similarly, one can use philosophical or theological theories to form models that enable him to see phenomena as examples of the model and to derive test results. Einstein's model of light as a series of particles impinging on a metal's surface was, in part, a philosophical model that likened photons to tiny billiard balls that bump into other balls (electrons) on the metal's surface.

James Clerk Maxwell (1831–1879) was a devout Christian and British physicist who proposed that light be pictured as a wave wherein electric and magnetic waves oscillate back and forth as the wave travels through space. Maxwell's field picture was derived metaphysically from his theological convictions of the Trinity and incarnation.[12] For example, in the Trinity, no one person can be viewed atomistically, but each must be viewed in relation to the others. If the world somehow reflects its Creator, Maxwell reasoned, then one can derive from theology a metaphysical picture of reality wherein relations among the parts of wholes, relations between wholes, and the wholes themselves are just as fundamental as are atomistic parts of those wholes. Thus, light should be viewed as a continuous field, a unity with different aspects (magnetic and electric) in relation.

The details of Maxwell's views need not concern us further. Suffice it to say that his theological concepts allowed him to derive metaphysical theories about aspects of the world. Justification of Maxwell's views of light is a multifaceted affair, relying on empirical data and the rationality of the conceptual scheme used. The important point is that Maxwell illustrates what is often the case in the history of science—one can derive scientific ideas from theological or philosophical models.

It is an example of the genetic fallacy in logic to fault some belief solely for where it came from. Where one gets an idea is a different issue from why one is justified in believing it. People who fault creationists for deriving their scientific hypotheses from the Bible or theological frameworks commit the genetic fallacy and are out of touch with the actual way science has been practiced repeatedly throughout its history. There is nothing wrong in principle with deriving the notion of a limit to evolutionary change from the theological concept of a kind or essence and trying to test that idea. If someone denies that this is scientific, then much of the history of science is unscientific as well.

#2: Scientific Problems

Often, though not always, scientists begin their research not with observations but with a problem to be explained. Scientists often attempt to answer *what* questions (what is the relationship between pressure and volume in a gas at constant temperature?), *why* questions (why do metals expand when heated?), and *how* questions (how does light dislodge an electron from the surface of a metal?). When scien-

12. See Thomas F. Torrance, *Christian Theology and Scientific Culture* (New York: Oxford University Press, 1981), pp. 1–72.

tists attempt to answer these and similar questions, they are attempting to solve scientific problems.

But what is a scientific problem? Larry Laudan, perhaps more than any other philosopher, has given detailed consideration to the nature and classification of scientific problems.[13] A scientific theory attempts to remove ambiguity, reduce irregularity to uniformity, and show that what happens in the world is somehow intelligible and expected. Scientists do this by solving problems. According to Laudan, there are two broad kinds of scientific problems: empirical and conceptual.

EMPIRICAL PROBLEMS

Empirical problems do not arise from some brute, bedrock confrontation with the world. Rather, they arise in the context of inquiry about the world and are theory-laden to a greater or lesser degree. Empirical problems are first-order problems about objects in some domain (e.g., chemical phenomena in acid/base reactions, inheritance phenomena in peas), and are, in general, anything about the natural world that strikes us as odd and in need of explanation. An empirical problem need not describe an actual state of affairs in the world; rather, it must be *thought* to describe such a state of affairs. For example, early nineteenth-century biologists, who believed in spontaneous generation, tried to solve the empirical problem of how meat could transmute into maggots.

There are three major types of empirical problems. First, there are unsolved problems. These are empirical problems that have not yet been adequately solved by any theory. These carry less weight than other kinds of problems for two reasons. First, since no available theory solves them, they count equally against all theories. Second, it is hard to know, sometimes, if the problem is one a particular theory ought to solve. For example, Laudan points out that from 1828 to 1850, Brownian motion (the random motion of microscopic particles, e.g., tiny particles visible through a microscope that dart about on the surface of a beaker of water) was unsolved. But different scientists classified it as a biological, chemical, optical, electrical, heat-theory, or mechanical problem. As long as it was unsolved, it could be ignored and classified as a problem in another branch of science.

More important are solved problems, that is, problems that at least

13. See Laudan, *Progress and Its Problems: Towards a Theory of Scientific Growth* (Berkeley: University of California Press, 1977), pp. 11–69. Thomas Kuhn has argued persuasively that different scientific theories or paradigms can have the effect of defining differently the methods, problems, and standards for what counts as science. Thus the very nature and classification of problems can themselves change, as does the very idea of what counts as science. See Thomas Kuhn, *The Structure of Scientific Revolutions*, 2d ed., enlarged (Chicago: University of Chicago Press, 1970), pp. 103–35.

one theory can solve, or several competing theories might solve in different ways. Sometimes a solution to a problem is not permanent. For example, wave theories of light solved problems associated with specifying the nature of light (e.g., why colors arise when light is passed through a prism), but new problems (e.g., the photoelectric effect requiring a particulate view of light) can make old solutions inadequate.

Also important for science are anomalous problems. These are problems that a particular theory has not solved, but that one (or more) of its rivals has solved. Thus, to count as an anomaly, there must be a rival theory that solves the problem. Anomalies are important, but not necessarily decisive, problems for a theory. One anomaly does not refute a theory; rather, it merely raises doubts against it. The strength of anomalous problems is a function of several things: How many of them are there? How well does the rival solve them? What is the quality of our data that these really are anomalies instead of poor instrument readings? How central is the anomaly to the theory in question?

CONCEPTUAL PROBLEMS

In addition to empirical problems, Laudan argues that scientists also try to solve conceptual problems. We have already discussed these in chapter 1, so we need not go into detail here. Conceptual problems have to do with the conceptual structure of the theory that exhibits them. Two broad kinds of conceptual problems exist.

First, there are internal conceptual problems. These arise when a theory exemplifies internal inconsistencies, vagueness, circularity, and so on. Laudan illustrates this with the example of Michael Faraday who, in attempting to describe electrical interaction without appealing to action-at-a-distance, merely reintroduced a shortened form of action-at-a-distance in his solution. Thus, the existence of such action remained a conceptual problem within his system.

Second, there are external conceptual problems. These problems conflict with some scientific theory, are rationally well established, and arise in fields outside of science like theology, philosophy, logic, mathematics, intrascientific debate (e.g., mechanistic views of a person versus Adlerian views), philosophical views about how science ought to be (e.g., Einstein's commitment to a deterministic, causal account of scientific explanation expressed in his famous statement "God does not play dice with the universe"), and world-view considerations. In assessing a particular scientific theory, these problems are relevant, even though they do not arise from observation or from science itself.

It should be obvious that scientists and philosophers differ over the role of each of these kinds of problems—unsolved, solved, and anomalous empirical problems and internal and external conceptual problems. They also differ over whether science merely describes phenomena (i.e., it answers *what* questions) or whether, in addition, it seeks explanations for the hidden structures, entities, and mechanisms behind observational phenomena (i.e., it also answers *how* and *why* questions). One's view of "the" scientific method will depend on her answer to these issues, and a variety of scientific methods will result from different answers.

#3: The Use of Scientific Ideas: Scientific Explanation

Scientists use laws and theories to explain the phenomena they investigate. In order to get an introductory grasp on scientific explanation, let us examine two main areas of focus: the nature of scientific explanation and the types of scientific explanation.

THE NATURE OF SCIENTIFIC EXPLANATION

What is it that scientists do when they use theories and laws to explain something? What are the important features of scientific explanation? There are three main schools of thought about the nature of scientific explanation: the covering-law model, the realist model, and the contextualist model.

The inferential or covering-law model. The first theory of scientific explanation has gone by several names: the inferential, covering-law, or deductive-nomological model, or simply deductivism. Its main advocates have been Carl G. Hempel and Ernest Nagel.[14] The covering-law model includes a picture of scientific explanation and a picture of the function of scientific theories.

14. Hempel, *Aspects of Scientific Explanation*, pp. 331–425; Ernest Nagel, *The Structure of Science* (Indianapolis: Hackett, 1979), pp. 15–78. Actually, the covering-law model has three different classes of scientific explanation: the deductive-nomological model (deductive explanations that include only universal laws in their explanans, e.g., all metal expands when heated, this is a metal, therefore this expands when heated), the deductive-statistical model (deductive explanations that include at least one statistical law in their explanans, e.g., 50% of radioactive substance x will decay in time y, this is z grams of substance x, therefore 50% of z will decay in time y), and the inductive-statistical model (inductive explanations that include at least one statistical law in their explanans, e.g., 90% of people who get penicillin recover, Jones got penicillin, therefore Jones recovered). I will focus only on the deductive-nomological variety of the covering-law model, though what is said about it could be applied to the other versions as well.

Let us look first at its picture of scientific explanation. According to this picture, the essence of scientific explanation is its form. To explain an event scientifically is to give a correct deductive (or inductive) argument for that event. Suppose we want to explain why a particular metal wire we are investigating conducts electricity. We could explain this fact scientifically as follows:

L_1: All metal conducts electricity.
C_1: This wire is a metal.   Explanans
_____   (Logical Deduction)
E: This metal wire conducts electricity.   Explanandum

The explanation has the form of a deductive argument where the thing to be explained (E)—called the explanandum—is a logical consequence of the explanans. The explanans includes at least one general law (e.g., L_1) with empirical content (i.e., capable of test by experiment or observation) and a statement of antecedent or initial conditions (C_1). The explanans "covers" the explanandum, that is, the general laws plus initial conditions allow one to deduce logically the thing to be explained. Thus, a good scientific explanation of some explanandum like E involves subsuming E under a general law that, given certain conditions, allows one to deduce E. Note that on the covering-law model there is a symmetry between prediction and explanation: the same structure or logical process applies to both. Explanation and prediction differ only in that what is to be explained has already occurred and what is predicted is still future. But every prediction counts as an explanation and vice versa.

Now let us look at the covering-law model's picture of scientific theories. To see this, consider another example. Suppose we are trying to explain why the second generation of peas that are crossbred occurs in the ratio of 3 green:1 yellow. The covering-law explanation would be this:

L_1: In the second generation the ratio of dominant to recessive characters is 3:1.
C_1: This is the second generation of peas.
C_2: The dominant character is green.
C_3: The recessive character is yellow.
_____
E: The second generation ratio is 3 green:1 yellow.

According to the covering-law model, the sole aim of a theory is to provide a basis for a deduction of the thing to be explained. Now, in most scientific theories, there are theoretical concepts: "gene," "dom-

inant gene," "recessive gene." Although it can be combined with scientific realism, the covering-law model tends to interpret a theoretical
concept nonrealistically, that is, as nothing more than a cipher, a
merely logical device used to summarize sense data and to deduce
empirical results. So understood, theoretical concepts do not refer to
real entities in the world. That simply is not their function in scientific
explanation.

The covering-law model of explanation does describe some of the
explanatory activity of scientists, for instance, explanation by appeal
to laws with empirical content, though even here it needs to be pointed
out again that such a form of explanation is used outside of science
(history, hermeneutics, religious experience).[15] Nevertheless, a number of fairly substantial criticisms have been leveled against it.[16] First,
conformity to the formal requirements of the covering-law model (i.e.,
that a general law and statement of antecedent conditions be present
in the explanans) is not sufficient for explanation, because for any
explanandum, an infinite number of sets of explanantes will entail it.
For example, "all wooden objects are conductors, this metal object is
wooden, therefore this metal object is a conductor" counts as an explanation on the covering-law model. In order to rule out examples of
this kind, the formal requirements of the covering-law model must be
supplemented by other requirements: that the explanans be true, simple, not yet falsified, and so forth. But these qualifications are prob-

15. On the other hand, attempts to extend the covering-law model into some domains are disastrous. For example, mental states like feeling anger, thirst, or pain,
and thinking about something are irreducibly mental phenomena that have a first-
person subjectivity (i.e., they are present to me as a subject in a way not reducible to
a third-person point of view outside me). Statements using such terms ("I am in pain")
are often used to report these states to others. However, physicalists like Paul Churchland use a covering-law model of explanation to reduce these states (Churchland
speaks of the associated terms, e.g., "pain," as opposed to the states they refer to) to
public, physical phenomena. Thus, according to Churchland, terms like "pain" are
part of a theory known as folk psychology (for example, I know Jones is in pain because
all people in pain tend to wince, Jones is wincing, therefore Jones is in pain). But
regardless of whether a covering-law model helps me understand when Jones is in
pain, that is not the same thing as Jones knowing that he himself is in pain, on the
one hand, and reporting that pain, on the other. When Churchland equates these first-
person experiences and their reports with the third-person covering-law explanations
embedded in folk psychology, he distorts the former. See Paul M. Churchland, *Matter
and Consciousness* (Cambridge, Mass.: MIT Press, 1984), pp. 56–61, 70–72.

16. See Harre, *Philosophies of Science*, pp. 53–58; Karel Lambert and Gordon Brittan, Jr., *An Introduction to the Philosophy of Science*, 3d ed. (Atascadero, Calif.: Ridgeview, 1987), pp. 22–25; or more briefly, Bynum, Browne, and Roy Porter, eds., *Dictionary of the History of Science* (Princeton, N.J.: Princeton University Press, 1981), s.v.
"explanation."

lematic in their own right. For example, several theories in the history of science provided explanations of phenomena for a long time even though we now believe them to be false, and this feature of the history of science is hard to square with the supplemented version of the covering-law model.

Second, it is hard to see how the covering-law model is really explanatory in itself. The statement "This x has F because all xs have F and this is an x" merely postpones explanation by inviting the question "Why do *any* xs have F in the first place?" An explanation of this latter question will use new concepts embedded in models of what x is like (e.g., metal conducts electricity because little entities called electrons are free in metals). So the covering-law model is not sufficient for an explanation.

Third, Wesley Salmon has pointed out that covering-law explanations are really just descriptions of what happens and how, but are not really explanations of why things happen.[17] He invites us to imagine that we have become one of Laplace's demons—we know all of nature's regularities and the precise state of the universe at any moment in such a way that we can predict any future state (or explain any past one) by inferring that state from general laws and boundary conditions. Still, Salmon argues, there would be something left out of our knowledge, namely, we would not know why things happen as they do. Such knowledge requires insight into the hidden causal mechanisms behind the phenomena of nature.

Fourth, the symmetry between explanation and prediction does not obtain in much of science. Thus, scientists could predict the atomic weights of uninvestigated elements in the periodic table long before they could explain why those weights were to be expected. On the other hand, scientists may some day be able to explain the extinction of the dinosaurs without being able to predict anything on the basis of their explanation.

Finally, the covering-law model cannot adequately distinguish between real causal laws and mere accidental generalizations. For example, from ancient times a lawful correlation between the position of the moon and the tides was well known, but a genuine prediction of the location of the tide during September based on the location of the moon would not be an explanation of the tide's location. Such an explanation would involve the notion of gravity, which causes the tide to do such-and-such. Again, for some time scientists could have noted a lawlike correlation between the heat of the sun and the growth of

17. Wesley C. Salmon, "Why Ask, 'Why?'? An Inquiry Concerning Scientific Explanation," reprinted in *Scientific Knowledge: Basic Issues in the Philosophy of Science*, ed. Janet A. Kourany (Belmont, Calif.: Wadsworth, 1987), pp. 51–64.

plants, but such a correlation would be merely accidental, since a modern theoretical explanation of plant growth uses the incident wave frequency of sunlight in the process of photosynthesis. But on the covering-law model, each explanation (the heat and the frequency of sunlight) would be appropriate, since both posit lawlike correlations.

Realist, causal models of scientific explanation: There are various realist models of scientific explanation, but let us look briefly at two: Wesley Salmon's causal-statistical model and Rom Harre's model.

Salmon's model of scientific explanation is somewhat complicated, so we can only attempt to give an overview of it here.[18] According to Salmon, one explains an event scientifically when one presents both the set of factors statistically relevant to it and the causal network underlying those regularities and the event itself. B is statistically relevant to A if the probability of A given B is greater than the probability of A without B. Wearing a hat is not statistically relevant to recovering from pneumonia, but receiving an injection of ampicillin is.

Now consider a scientific explanation of why some soldier S got leukemia. Such an explanation involves (a) giving the set of statistically relevant factors (e.g., soldier S was 2 kilometers from the hypocenter of the atomic explosion and unsheltered, the explosion was a one-megaton explosion), (b) giving statistical regularities relevant to (a) (e.g., the probability of a person's getting leukemia given that he was at a distance of 2 kilometers from a one-megaton atomic explosion is 1/1000, a number much higher than the general average), and (c) listing the causal processes and interactions underlying the statistical regularities (e.g., a causal process like the fact that high radiation released in a nuclear explosion traverses the space between the explosion and soldier S, strikes the cells of S, interacts with those cells, is absorbed by them, and initiates a process leading to leukemia).

The important point for Salmon is that a scientific explanation not only subsumes events under laws but also offers understanding of the causal network underlying the events.[19]

Rom Harre is another realist who has analyzed the nature of scientific explanation.[20] According to Harre, scientific explanation is not essentially a matter of mathematics and quantification. For example, the concept of a virus explains certain pathologies without making essential reference to mathematical considerations. Frequently, a scientific explanation postulates some unseen entity, the existence of

18. My discussion is heavily indebted to Lambert and Brittan, *Introduction*, pp. 26–34. See also Achinstein, *Nature of Explanation*, pp. 9–11, 173–76.
19. For criticisms of Salmon, see Lambert and Brittan, *Introduction*, pp. 31–34.
20. See Harre, *Philosophies of Science*, pp. 168–83.

which is postulated on the basis of some sort of analogy, to explain a range of phenomena. The laws of the theory do not make essential reference to mathematical expression; neither do they necessarily fit together into a formal, logical system. For example, Harre points out that our theory of human behavior is a grab bag of principles united by virtue of the fact that they all are concerned with the same subject matter, that is, the behavior of individuals.

Further, a scientific explanation consists in describing the mechanism that produces the phenomena being explained. Often, we see patterns and regularities in nature and, by the use of analogy, postulate theoretical entities and causal mechanisms of which we have no independent access apart from the phenomena to be explained and the theory we have formulated. We often work with only one term in the analogy. For example, we liken a molecule to a particle, say a billiard ball, in motion, but we cannot examine the molecule directly to see how far the analogy can be pressed accurately. We are entitled to affirm all and only those properties of the theoretical entity thought necessary to explain the phenomena in question. In order to do this we use an iconic model, that is, a real model wherein certain things and processes are modeled on things that we know but are models of the unseen things and processes thought responsible for the thing in question.

An example may be helpful. The kinetic model of a gas pictures the relationship between the temperature, volume, and pressure of a gas according to a billiard-ball model of gas molecules. On that model, gas molecules are likened to tiny billiard balls that undergo elastic collisions, travel through space, bounce off of container walls, and so forth. There is an analogy between the iconic model (billiard balls) and the unseen particles and their behaviors. We ascribe to the latter just what is necessary to explain temperature, pressure, and volume phenomena of gases. Further, we have no independent access to these molecular particles apart from our theory of them. So we cannot declare any negative analogy (respects of difference) between the molecules and the billiard balls since the molecules are only invested with those positive likenesses that the phenomena demand. Nevertheless, to explain a phenomenon is to devise a model by analogy that explains the causal mechanisms responsible for the phenomena in question.

I do not wish to criticize realist interpretations of scientific explanation here. We will discuss in later chapters the debate between realism and antirealism in the philosophy of science. It does seem, however, that realist accounts of scientific explanation capture much, but not all, of what happens in scientific explanation. For it sometimes happens that scientists cannot offer or agree on a causal mechanism of some phenomenon. Or, in the case of quantum phenomena, they may not even think that there is a causal mechanism to be discovered. In these cases, scientists still can offer some kind of explanation of the phenomena in question, perhaps along the lines of the covering-law approach.

Before moving to the third main school of thought regarding the nature of scientific explanation, an application to theology should be drawn from the realist model, especially as it is formulated by Harre. In his excellent work *The Road of Science and the Ways to God*, Catholic priest and realist philosopher of science Stanley L. Jaki argues over and over that the type of explanation a realist envisages for science is the same kind of explanation theists use in arguing for God from various phenomena in the world (e.g., its existence, its various forms of design, the existence of the human mind, and so forth). According to Jaki, both the scientists and the theologian make a bold jump of the intellect beyond data to an intelligible theoretical entity postulated to explain the data.

In my opinion, Jaki is quite correct. In natural theology, the student of the various aspects of the universe and religious experience postulates a cause that is at least partially understandable or intelligible to the human intellect; is, strictly speaking, beyond those data themselves and serves as the ground and explanation of those data; is given those attributes deemed necessary to explain the data; and makes no essential reference to quantitative, formal, or mathematical expressions, although they can be used on occasion.[21] This same point has been made more recently by Princeton philosopher and antirealist Bas C. van Fraassen. Van Fraassen argues that if one accepts a realist account of scientific explanation, then he must also accept classical examples of natural theological explanations, for example, Thomas Aquinas's famous five ways to prove God's existence, since both sorts of explanations have the same structure.[22] It would seem that if a realist account of scientific explanation is accepted, then it becomes difficult to sustain a dichotomy between scientific methodology and theological methodology, at least as far as explanation is concerned.

21. See Stanley Jaki, *The Road of Science and the Ways to God* (Chicago: University of Chicago Press, 1978). See also Edward L. Schoen, *Religious Explanations: A Model from the Sciences* (Durham, N.C.: Duke University Press, 1985), pp. 51–148.

22. See Bas C. van Fraassen, *Scientific Image* (Oxford: Clarendon Press, 1980), pp. 204–15. For a criticism of van Fraassen (which does not break but only changes the comparison between theology and science), see Richard Boyd, "Lex Orandi Est Lex Credendi," in *Images of Science: Essays on Realism and Empiricism, with a Reply from Bas C. van Fraassen*, ed. Paul M. Churchland and Clifford Hooker (Chicago: University of Chicago Press, 1985), pp. 32–33.

A contextualist or pragmatist account of scientific explanation. The third school of thought is a rather loosely associated cluster of thinkers, among them van Fraassen, Kuhn, and Laudan, who agree that the covering-law and realist models of scientific explanation are inadequate. In chapter 5 we will examine in some detail these men's views of science. For now we can content ourselves with three brief observations.

First, contextualists, or pragmatists, believe that scientific explanation attempts to answer *why* questions (Why do marsupials live only in certain regions? Why can gravity act at a distance?) that arise when we entertain empirical or conceptual problems about the world. Second, conceptualists believe that there is no particular way to characterize scientific explanation, since there are several types of answers scientists give to the *why* questions generated from the puzzles they consider—some quantifiable, some not, some appealing to singular phenonema (the Big Bang), some to repeatable phenomena, some referring to unobservable entities, some not, some using probabilities, some not, and so forth. Scientific explanations are, at best, a cluster or family of different kinds of explanations. Third, contextualists hold that there is no clear-cut line of demarcation between scientific and nonscientific explanation, no particular form and no particular sort of information that all and only scientific explanations embody.

THE KINDS OF SCIENTIFIC EXPLANATIONS

In addition to the debate over different forms of scientific explanation, there are also several different kinds of scientific explanations.

Compositional or structural explanation. Here the properties of an object are explained in terms of the properties or structural relations of its parts. For example, the properties of the chemical elements are explained in terms of the properties and relationships of the neutrons, electrons, and protons in the elements.

Evolutionary or historical explanation. Here the properties or some other aspects of an object are explained in terms of the temporal development and history of the object and its ancestors. Evolutionary theory regarding living organisms is an example of this kind of explanation, as is the evolutionary development of a star.

Functional explanation. Here the capacities of an object are explained in terms of the function it plays in some system. Thus, the nature of the human heart is explained by reference to the role the heart plays in the circulatory system. Such explanations take the following form: "The function of x is to do y."

Transitional explanation. Here a change of state in an object—a transition from one state to another—is explained in terms of some disturbance in the object and the state of the object at the time the disturbance took place. For example, the change in the motion of an object is explained in terms of its initial motion and the forces acting on it.

Intentional explanation. Though controversial, some psychologists and biologists explain the behavior of an organism in terms of the beliefs, desires, fears, intentions, and so forth, of the organism. An example would be the explanation of anger or aggression in a person in terms of his fear of a loss of self-esteem.

#4: Scientific Experiment

Scientists (as well as nonscientists) make observations and do experiments.[23] An observation occurs when an observer stands outside the course of events in which she is interested. She does not manipulate the process, she just waits and watches. But even here, the observer usually has to have in place already a well-worked-out theory in order to know what to look for, what counts as significant, and what counts as real seeing as opposed to observational error.

An experimenter is in a different relationship to nature. She intervenes in the course of events, seeks to focus on specific parameters of some phenomenon in question, and sets up experiments to control variables thought relevant to measure the phenomenon under investigation. An independent variable is a factor in the setup that one can manipulate directly; a dependent variable is a factor that is affected by changes in the independent variables. The dependent/independent classification is, in turn, usually dependent not on observation, at least from the start, but on the background theory of the thing being investigated.

There are at least three kinds of instruments scientists use in making observations and doing experiments. First, there are instruments used in making measurements (e.g., clocks, meters, and thermometers). Some of these are self-measurers, that is, they are examples of the thing being measured. A meter rod has length and its length is used to measure length. Some of these are non-self-measurers, that is, they use effects of some phenomenon to measure the cause. Thermometers directly measure temperature, but it is assumed that this effect

23. See Harre, *Great Scientific Experiments*, pp. 1–22; Robert John Ackermann, *Data, Instruments, and Theory: A Dialectical Approach to Understanding Science* (Princeton, N.J.: Princeton University Press, 1985), pp. 149–64.

is correlated with temperature as its cause. Observations from each kind of instrument, especially the latter, depend on background assumptions from physical theory (e.g., that temperature uniformly causes expansion, that the ruler does not shrink as it is shifted from place to place along the side of the measured object, and so forth).

Second, there are instruments that extend our senses (e.g., microscopes, telescopes, and amplifiers). Observations from these instruments depend on background assumptions as well, especially the physics of light and theories about how light interacts with objects. For example, through telescopes astronomers observed that galaxies that are very far away appear redder than similar objects taken to be closer. How should they interpret this? Are these bodies the same distance away as bluer objects but emitting redder light, or are they farther away, receding from us more rapidly, and emitting light that is red shifted? Scientists have opted for the latter, but their choice was interpreted, not observed through the telescope.

Third, there are instruments that isolate phenomena to allow them to be studied independently from their environment. Outside influences are either eliminated or controlled and kept constant. Observations based on this type of instrument assume that the phenomenon in question does not change significantly when it is isolated from these external features.

As we have seen, all three types of instruments (and experiments or observations based on them) depend on a variety of background assumptions. These assumptions may or may not be beyond reasonable doubt, but in any case, an experiment or observation that uses instruments is not simply a matter of seeing if the pointer is or is not at 3. If something unanticipated is observed that tends to falsify some theory, one can always assume that the instrument functioned poorly or call into question some of the background theory assumed in using the instrument. Thus, the use of instruments complicates the role of falsification and verification in science. This does not mean that instruments complicate this role beyond recognition; nevertheless, their use illustrates the fact that scientists often test groups of theories simultaneously (the specific theory being tested, background theories about the instrument, etc.), not just single theories.

#5: Testing Scientific Ideas: Scientific Confirmation

Before a scientific theory is regarded as well established, it must prove itself by being tested. But how do scientists test and confirm a proposed theory? People commonly think scientists test and evaluate theories simply by the results of observation and experiment. But the history of science itself amply shows that this is not the case. Testing and evaluating a scientific theory are complicated affairs. One reason is that scientists often make claims beyond observational evidence by postulating unseen theoretical entities. In these cases, several different theories will always be compatible with the observational evidence. Further, a particular theory thought to be confirmed could really be false but, nevertheless, reasonably embraced given the current state of the evidence.

These and other ambiguities about theory confirmation in science have given rise to several competing schools of thought about the nature of scientific theory confirmation and, thus, about the nature of scientific methodology.[24] In this section we will explore three different understandings of scientific theory confirmation—conventionalism and pragmatism, falsificationism, and justificationism.[25]

CONVENTIONALISM AND PRAGMATISM

There have been several proponents of this first school of thought, including the physicist Pierre Duhem, physicist and philosopher Thomas Kuhn, and philosophers Paul Feyerabend, W. V. O. Quine, and Larry Laudan. Suppose two theories are being compared and tested against experimental data in such a way that the data appear to confirm one of the theories better than the other. According to conventionalists and pragmatists, the data really do no such thing. Data do not show that one theory is false and the other probably or even possibly true. With sufficient adjustment, both theories could have accommodated all the experimental results and it is difficult, if not impossible, to state general criteria for when such adjustments are inappropriate. This is because theory assessment, as evidenced by the history of science, is a multifaceted affair, and theories simply are not chosen on empirical grounds. Facts can be accommodated by any conceptual framework, and these frameworks are chosen either by pure convention or by rational factors (e.g., simplicity) that go well beyond empirical observation of data. Further, theories are tested not in isolation but in conjunction with a whole system of auxiliary theories about instruments, or other physical phenomena besides the ones under investigation.

We will look at two representatives of this school in chapter 5, Kuhn

24. Kourany lists six different schools of thought regarding scientific confirmation. See Kourany, *Scientific Knowledge*, pp. 112–21.

25. I have listed conventionalism and pragmatism together for the purposes of our discussion since they are united in rejecting the other two approaches. However, conventionalists and pragmatists do differ, and some of these differences will be explored in chapter 5.

and Laudan, so we do not need to develop our discussion further here except to draw out three final points. First, several nonempirical factors are determinative in assessing a theory—simplicity, predictive ability, explanatory scope, accuracy, consistency and clarity, fruitfulness for guiding new research, and the presence or absence of a rival explanation. Second, members of this school argue that they have the history of science on their side; that is, if one studies how science actually is and has been practiced, then the crucial-experiment model— two rival theories are decisively tested by some crucial experiment yielding unambiguous data—shows itself naive and a misrepresentation of real science, even if most modern scientists mistakenly believe it is accurate. Third, conventionalists and pragmatists emphasize that there is no clear line of demarcation between science and nonscience regarding methods of confirmation and, furthermore, that philosophical and theological consideration legitimately figure into scientific theory choice.

FALSIFICATIONISM

The main advocate of falsificationism has been Karl Popper.[26] According to Popper, theories of scientific theory testing have placed their emphasis on the positive confirmation of a theory by experimental data. But this emphasis is misplaced. Rather, evidence is valuable when it is seen as having the potential not to verify a scientific theory but to refute or falsify it. What some take as positive evidence for a theory experimentally is really a failure to falsify the theory. Such a failure, says Popper, does not even make the theory probably true; it only shows that the theory is possibly true.

Popper's reasons for falsificationism include the following consideration: There is an important difference between confirming and falsifying hypotheses that are universal generalizations, for example, "all ravens are black." No amount of confirming instances, no matter how many black ravens we discover, shows that the generalization is true, for we could always find in the future a white raven. So positive confirmation is always indecisive. Induction—inference based on many observations—is a myth. By contrast, a single disconfirming instance—a single nonblack raven—can serve to falsify the universal generalization. When we add to this observation the fact that only a small fraction of the potentially infinite observational consequences of a theory are ever observed, and these are themselves taken from a minuscule part of the universe during a very small period of its total

26. Cf. Popper, *Conjectures and Refutations: The Growth of Scientific Knowledge* (New York: Harper and Row, 1963), pp. 33–65.

duration, then we realize that even a large number of positive instances of a theory do not even make it probably true, but only possibly true.

So, Popper concludes, scientists should proceed not by offering conservative theories that are merely broad enough to cover the present data but by bold, risky conjectures—jumps to conclusions often taken after only one observation—with a lot of empirical content (i.e., observational consequences) that, in principle, can be easily falsified. A genuine test of such a theory counts for the theory in the sense that it was an unsuccessful attempt to falsify it.

At this point a problem arises for Popper. According to him, a scientific theory is successful if it has so far stood unfalsified. The theory that the combustion of a metal involves oxidation (adding oxygen to the metal) has not been falsified so far. But neither has the theory that the 298th star from the sun is made of green cheese. But surely there is a difference between these two theories in that the former has a higher scientific status than the latter even though neither has been falsified. The difference between them, says Popper, is that the former has been corroborated while the latter has not. A theory is corroborated just in case it has been tested several times and many attempted refutations have been undertaken. Such theories are highly corroborated (severely tested and unfalsified) even though we have no idea how probable they are. The oxidation theory has been corroborated, but the theory about the 298th star from the sun has not.

Finally, Popper states a clear line of demarcation between scientific and nonscientific theories—the former are falsifiable but the latter are not. This does not mean, for Popper, that nonscientific theories are meaningless, unreasonable, or false; it only means that they are nonscientific.

Several objections have been raised against falsificationism. First, Popper has a simplistic view of how data falsifies a theory. As mentioned earlier, one can always readjust a theory in the face of negative evidence, and no clear criteria exist for telling when such readjustments are inappropriate. Further, even if the evidence is taken as falsificatory, it may be one of the other auxiliary theories that is false (remember, theories are tested in groups) and not the central theory. Finally, there is no set of rejection procedures for a theory even if the evidence is taken as tending to falsify it. On balance, the positive reasons for accepting the theory may far outweigh the negative impact of discordant data.

As an example of this last point, astronomers in the middle of the nineteenth century were faced with the fact that irregularities in the motion of Uranus could not be accounted for by Newton's laws of motion and gravitation in terms of the gravity of the sun and other known planets. At least three options were available to them. First, they could conclude that Newton's theories had been falsified. (This option would seem to be implied by Popper.) Second, they could restrict the range of applicability of Newton's law of gravitation and make it a satisfactory explanation of planetary motion out to a distance from the sun just short of Uranus. Third, they could postulate the existence of an unknown planet with sufficient mass and location to explain the irregularities. Scientists who adopted the third strategy won the debate when Neptune was discovered. But prior to that discovery, none of the three options was dictated by the evidence, and Popper's view would seem to have favored the first option.

Second, Popper's doctrine of corroboration seems to be just another name for confirmation. When a positive observation obtains for some theory and we say that it confirms the theory, we do not mean that the theory is conclusively proven; we only mean that it has been given some positive support. A theory that has been tested repeatedly and so far has been unfalsified would, by that very fact, seem to accrue positive confirmation, even though it could still turn out to be false.

Finally, Popper's line of demarcation between science and non-science—especially metaphysics—is not widely accepted. We have already discussed the reasons for this rejection in this and the previous chapter. Simply put, scientific theories themselves include metaphysical statements (about the properties and relations of unseen theoretical entities, e.g., carbon is a *substance* with such-and-such *properties* and *dispositions*) and are not straightforwardly falsifiable as Popper claims. Rather, scientific theories are abandoned not only because some observations seem inconsistent with them but also due to other factors (e.g., simplicity, lack of clarity). Furthermore, philosophical and theological claims are often falsifiable (e.g., the resurrection of Jesus, certain predicted results of certain spiritual practices).

JUSTIFICATIONISM

It has been held traditionally that confirmation of scientific laws or theories comes by way of positive instances of those laws or theories. While there are several different versions of justificationism, all of them agree that positive instances of a law or theory not only fail to falsify it but actually serve as positive supports. A scientific theory is not totally proven by positive instances, but those instances nonetheless tend to confirm the theory they instantiate, and confirmation is, if not totally at least in large measure, a matter of whether a theory has a large number of positive instances. For example, the larger the actual number of cases where temperature, pressure, and volume measurements of gases exemplify the ideal gas law (PV = nRT), the more that law is confirmed. We will look at two different versions of justificationism—the positive instance account and the Bayesian account.[27]

The positive-instance view of confirmation. The major advocate of this view has been Carl Hempel.[28] According to this view, the direct evidence E for a hypothesis H is the set of positive and negative instances of H. Suppose a scientist tried to confirm the generalization "All A are B." All objects that have both A and B are positive instances of the generalization. If an object is A but not B, then it is a negative instance of the generalization, and if an object is not A to begin with then it is neither a positive nor a negative instance. According to the positive-instance view, the direct evidence for H is the set of positive instances of H and the direct evidence against H is the set of negative instances of H. H is confirmed (i.e., supported) by its positive instances and disconfirmed by its negative instances.

The idea behind the positive-instance account of scientific confirmation is induction by simple enumeration. Here one supports a generalization by simply enumerating (listing) the instances of the generalization, and this list serves as the premises in an inductive argument with the generalization as the conclusion. This can be illustrated as follows:

$Raven_1$ is black.
$Raven_2$ is black.
$Raven_3$ is black.

.

.

.

All ravens are black.

The basic idea here is that to test a generalization one merely looks at samples of the generalization, both positive and negative.

Although the positive-instance view has considerable intuitive appeal, it has not received widespread acceptance by philosophers and historians of science because of objections raised against it. Three of them will be mentioned here.

First, historians and philosophers of science point out that the great majority of scientific debates about the adequacy of hypotheses have not been resolved merely by pointing to which hypothesis has the most positive instances. Several other factors—simplicity, clarity, novel

27. See Lambert and Brittan, *Introduction*, pp. 75–99.
28. See Hempel, *Aspects of Scientific Explanation*, pp. 3–96.

explanations in new domains, and so forth—all play a role and can be more determinative in acceptance of a hypothesis than the mere number of positive instances that can be cited in its favor. Positive instances do play a role, but they are only one factor in a complex process of hypothesis evaluation that involves several factors. These factors are weighed in different ways, and only by looking at the particular case in question can one, at least in principle, decide which factors are key. In general, positive instances play the most important role, though not the only one, in areas of science where statistical generalizations are sought (e.g., what percentage of people vaccinated with a certain drug lose symptom y).

Second, Harvard philosopher Nelson Goodman's "new riddle of induction" paradox,[29] if successful, raises difficulties about what counts as a confirmable hypothesis in the first place. Goodman argues that not every generalization is supported by its positive instances. Consider the following two generalizations:

(1) All emeralds are green.
(2) All emeralds are grue.

X is grue if and only if either x is examined before time t and x is green, or x is not examined before time t and x is blue. Suppose t is right now. All the positive instances of green emeralds up to the present moment would confirm both (1) and (2). Which generalization should we project into the future to predict the color of emeralds tomorrow? Positive instances alone do not settle this issue. We believe that (1) is a lawlike generalization and (2) is an accidental generalization with a contrived predicate. This belief is not based on positive instances, says Goodman, but on which predicate is more deeply entrenched in our language: we have used "green" to predict emeralds in the past, not "grue." Further, for any finite number of positive instances of green emeralds, there is a potentially infinite number of alternative predicates that could be confirmed by those instances. So instances alone do not settle the confirmation issue.

Finally, Hempel himself raised a problem known as the "raven paradox" that points out difficulties in the very notion of a positive instance.[30] If one were seeking to confirm generalization H, "All ravens are black," then one would, presumably, look around for ravens. These

29. Nelson Goodman, *Fact, Fiction, and Forecast* (Indianapolis: Bobbs-Merrill, 1965). For a brief overview of Goodman's riddle of induction, see Losee, *Historical Introduction*, pp. 195–97.

30. See Hempel, *Aspects of Scientific Explanation*, pp. 14–20.

would seem to constitute the appropriate class of instances of the hypothesis. But H is logically equivalent (by contraposition where one takes a statement and forms a logically equivalent one by switching and negating the subject and predicate terms in a categorical proposition) to H', "All nonblack things are nonravens." Remember, the positive-instance view of confirmation holds that confirmation consists in a purely formal relation between the enumerated instances of a generalization and the inductively derived generalization itself, as seen in our listing of the black ravens earlier. But a black raven is logically equivalent to a nonblack nonraven, so if the former inductively confirms the generalization "all ravens are black," so does a nonblack nonraven. An example of the latter is a white tennis shoe. Thus, a white tennis shoe would be a positive confirming instance of the hypothesis H, and a bird watcher could confirm H without ever leaving her house if she accumulated several nonblack nonravens in her home!

But surely a black raven is more important for H than is a white tennis shoe, and since the positive-instance view cannot capture the difference, it must be judged inadequate as a total theory of scientific confirmation. This is not to say that positive instances do not play a key role in confirmation. But it is to say that that role is complicated and not captured by the positive-instance view.

The Bayesian account of confirmation. A number of thinkers have tried to spell out scientific confirmation by using what is called Bayes' theorem, named after its discoverer, Thomas Bayes (1702–1761). In order to understand Bayes' theorem, we first must define what is called conditional probability. We call $P(T/E)$ the conditional probability of T given E, that is, the probability that T is true given that E occurs. Here T could be some scientific theory that we are attempting to confirm and E some evidence relevant to T's confirmation.

If evidence E confirms some theory T, then $P(T/E)$ is greater than $P(T)$, that is, $P(T/E)/P(T)$ is greater than 1 (if something has no probability, e.g., x is a square circle, its probability is 0; if x must be true, e.g., x is either real or not real, then its probability is 1). In other words, if some piece of evidence confirms some theory, then the probability of that theory being true given that the evidence obtains—$P(T/E)$—will be greater than the probability of that theory being true without evidence E—$P(T)$.

Bayes' theorem can be stated as follows:

$$P(T/E) = \frac{P(E/T) \times P(T)}{P(E)}$$

P(T/E)   = the posterior probability of T given evidence E.
P(E/T)   = the likelihood of E given T (i.e., how strong is the connection
              between T and E; does acceptance of T entail or just make likely
              the occurrence of E?).
P(T)   = the prior probability that T is true apart from evidence E. This is
              the prior probability of T.
P(E)   = the probability that E will obtain apart from acceptance of T,
              called the expectedness of E.

Bayes' theorem states that the probability of T being true in light
of evidence E is equal to the prior plausibility of T being true apart
from evidence E, multiplied times the likelihood that E is predicted
to occur if one accepts T, divided by the probability that E is expected
to occur without T being true at all. Scholars differ over what the
numerical values of probability mean when they are assigned to the
four terms in Bayes' theorem, but many hold that they are subjectively
assigned. This does not mean that they are not objective but merely
reflect someone's bias. Rather, it means that the probabilities represent someone's degree of belief or educated guess at the probability in
question.

An example may help to clarify Bayes' theorem. Suppose a robbery
has just been committed. T is the hypothesis that Smith did the crime,
and E is a body of circumstantial evidence: that witnesses saw Smith
around the scene of the crime the evening it happened, that Smith
had burglar tools in his trunk, and so forth. How much does E (the
evidence) confirm T (the hypothesis that Smith did the crime)? The
answer is a function not only of P(T/E) but of P(T/E)/P(T). Or, put
somewhat differently, the degree of support E gives to T = P(T/E) −
P(T). If the probability of T given evidence E is greater than the probability of T without evidence E, then E confirms (gives support to) T.
For example, if P(T) was 1, that is, it was certain that Smith did it
(fifty police officers saw him do it and captured him during the robbery), then the burglar tools in his car would add nothing to belief in
his guilt—we would already be certain of it.

Returning to our original example, suppose we determined P(T),
the probability that Smith did it apart from evidence E, say by noting
that he has a long criminal record, he often does robberies in this
vicinity, he likes to steal stereos, and a stereo was taken in this crime.
Suppose further that we know of no other criminal working this area
of town in the last eighteen months. We might guess that P(T) apart
from E is slightly greater than 50%, say .6. What about P(E/T)? If
Smith really did it, would we expect eyewitnesses to see him in the
vicinity and to find tools in his car? It would seem so, for if he is really

guilty, then we would expect not to find good testimony that he was elsewhere at the time of the crime and we would expect to find evidence in the vehicle he used to get to the scene of the crime. So $P(E/T)$ would be, say, .9.

What about the probability of the evidence E obtaining without T being assumed? In other words, what is the chance that Smith would be seen walking at the scene during the time of the crime and have burglar tools in his car if he did not do it? The probability of one state of affairs (e.g., that there is evidence which would convict Smith) is greater than or equal to the probability of two states of affairs (e.g., that such evidence exists and that Smith in fact committed the crime). In other words, $P(E) \geq P(E \text{ and } T)$. Now $P(E \text{ and } T) = P(E/T) \times P(T) = .9 \times .6 = .54$. So let us assign a value of .7 to $P(E)$.

According to Bayes' theorem

$$P(T/E) = \frac{.9 \times .6}{.7}$$
$$= .77$$

Further, $P(T/E) - P(T) = .77 - .6 = .17$. This is a positive number, so evidence E does in fact confirm T.

In science, when some experimental or observational evidence E obtains, its degree of confirmation of some theory T can be estimated in roughly the same way, according to advocates of the Bayesian theory of confirmation. Some scientists and philosophers accept this account of confirmation of scientific theories. Nevertheless, some have rejected it. Their reasons are rather technical, but a few can be listed here.

First, there is the problem of prior probabilities. Bayes' theorem provides a way to determine a *new* probability—$P(T/E)$—based on old or *prior* probabilities (the other three in the theorem). But where do these prior probabilities come from? We somehow must be able simply to assign them a value, but scholars differ over how to interpret these probabilities, and different people will assign them different values. For example, what degree of confirmation does the discovery of information and order in DNA confer upon the hypothesis of design or scientific creationism? That will depend on the value assigned to the prior probabilities of theistic design itself, the likelihood of finding information without a mind behind it, and so forth. But different scholars will approach these issues differently.

Second, probability arguments like Bayes' theorem have not been widely used in science, now or in the past, and actual scientific confirmation involves several factors apart from specific experimental evidence, factors that come together in ways that cannot be specified in a step-by-step algorithm for the rationality of theory choice. Bayes' theorem is overly simplistic, is at best only a small part of theory confirmation, and doesn't square with most cases of actual scientific practice.

Third, there is the problem of what is called old evidence. Many theories are accepted not because they give new predictions that are later verified but because they account more successfully for evidence that has been known for a long time. But in these cases, the evidence E is certain so P(E) is 1. Further, P(E/T) is 1, since the theory was formed to account for or guarantee that E would be explained. If these are put into Bayes' theorem, then P(T/E) equals P(T) multiplied and divided by 1. When one obtains the degree of confirmation E gives to T by subtracting P(T) from P(T/E), one gets zero. But surely the old evidence does confirm T—in spite of what Bayes' theorem says—because T was introduced in light of E and to explain E.

Finally, Bayes' theorem fails to account for how evidence can confirm a new theory that, in light of a well-accepted competitor, has a very low probability of being true, that is, P(T) for the proposed new theory is extremely low. In this case a positive instance of evidence E would confer very little support to T according to Bayes' theorem because of the low value of P(T) in the numerator. For these and other reasons several scholars have rejected the Bayesian account of confirmation.

#6: The Nature of Scientific Ideas: Laws and Theories

So far we have been using the terms *law* and *theory* in our discussion of scientific methodology without defining them. Unfortunately, there is no universally accepted definition of either term, and one's understanding of the nature of scientific law and theory, as well as of their relationship, depends in part on one's view about the realism/antirealism debate. Since this debate will be the topic of chapters 4 and 5, we need not go into detail here. But a few points can be highlighted that illustrate some of the different issues involved in understanding what scientific laws and theories are supposed to be.

First, there are three basic ways to distinguish a law from a theory. One way is to hold that a theory is roughly a hypothesis and, if it becomes well confirmed, can graduate to the status of a law. On this view, the only difference between the two is that a theory is held tentatively and a law is held firmly, that is, the difference lies in their relative degree of epistemological strength. While this way of speaking is fairly popular, it is the least helpful for understanding the nature of scientific methodology and, thus, is not widely used among philosophers of science.

A second way to distinguish a law from a theory focuses on their relative degrees of generality—a theory is broader in scope than is a law. For example, Kepler's laws of planetary motion or Galileo's law of free fall ($s = 16t^2$ where s is distance and t is time) only hold for a limited range of phenomena when compared to Newton's laws of motion, which apply not only to the same range as Kepler's and Galileo's laws but also to other heavenly bodies and other forms of motion. Here, theories are broader than laws and can incorporate them. Hence, distinguishing laws from theories this way, we would call Newton's laws of motion theories instead of laws.

A third way to distinguish theories from laws is embraced especially by those who hold to some form of scientific realism. On this view, laws merely *describe* the lawlike regularities that are observed in nature, and theories *explain* those regularities by offering a model for the theoretical entities, structures, and processes thought responsible for those regularities. For example, the ideal gas equation, $PV = nRT$, is a law, and the kinetic gas theory—gases are like swarms of little billiard balls that undergo elastic collisions with each other and with the container walls within which they are housed—is a theory or model that explains why gases behave as they do and so why the ideal gas equation holds true. Similarly, Mendel's laws of heredity are scientific laws, and the scientific picture of genes, phenotypes, and genotypes is part of the theory that explains those laws.

A second observation about scientific laws and theories involves the different ways scientific laws can be classified. First, "law" can refer to a linguistic or conceptual entity. In this sense a law is something that a scientist can discover at a point in time; he can write it on a sheet of paper or have it in his mind. On the other hand, "law" can refer to a real disposition (e.g., hydrogen atoms will become positively charged if they lose an electron) or relationship (e.g., an increase in the temperature of a gas will increase its pressure at constant volume) that obtains in the world itself, independently of anyone's discovery of it. In the former sense of law, it is appropriate to say that a law is an invention or a discovery of the human mind and doesn't cause anything to happen since it is merely our description of the world. In the latter sense, a law does cause something to happen. When some state of affairs obtains within some system (e.g., heat is added to a gas or two smooth peas are crossed), then the lawlike relations or dispositions of the system do cause the system to unfold in certain ways.

So whether a law can be said to cause something depends on how one is using the word *law.*[31]

In addition, laws can be statistical ("The probability of an atom of U^{238} decaying during this time period is .5," "The probability of dying from such-and-such a disease is .8") or nonstatistical ("All copper expands when heated"). Further, good scientific laws are not always true (e.g., the ideal gas law doesn't hold for any gas in the real world but only for idealized gases), but it is a helpful approximation even though it is strictly false.

Third, scientific theories can be classified according to various types. Frederick Suppe lists the following kinds:[32] theories that are intrinsically mathematical and involve a description of a state-transition mechanism (e.g., qualitative analysis in chemistry, laws of motion); theories like the first but that involve no description of a state-transition (e.g., classical optics, theories of chemical equilibrium); stochastic (involving a random or probabilistic process) theories, which are mathematicized or quantized but for which the basis for quantizing is counting and not measuring, and which do not usually involve a change of state (e.g., genetics, demography, epidemiology); qualitative theories that concern themselves with such problems as recognition and meaningful classification, theories that organize our observations and facilitate predictions (e.g., "What is a social action?" "What is a symbiotic relationship?" or cases of qualitative analysis in chemistry); taxonomic theories that are similar to qualitative theories but have a stipulative feature (e.g., biological classification, classification of natural languages); historical theories that describe a single event rather than a class of events by giving genetic reconstructions of the event (e.g., continental drift theory); theories in the social sciences that seek to impart an intuitive understanding of social behavior, institutions, political systems, and so on (e.g., Freud's psychology).

Finally, theories are interpreted differently depending on one's attitude about scientific realism.[33] Theories can be viewed as formal languages that are useful tools for drawing inferences and covering facts to be explained (instrumentalists), as summaries of sense data (positivists or phenomenalists) or of laboratory operations (operationalists), as rational structures imposed by the mind on the flux of sen-

31. For an analysis of several different ways of understanding scientific laws, see Del Ratzsch, "Nomo(theo)logical Necessity," *Faith and Philosophy* 4 (October 1987): 383–402.

32. See Frederick Suppe, ed., *The Structure of Scientific Theories*, 2d ed. (Urbana, Ill.: University of Illinois Press, 1977), pp. 123–24.

33. Cf. Ian Barbour, *Issues in Science and Religion* (New York: Harper and Row, 1966), pp. 162–71.

sory data (neo-Kantian idealists), or as true (or appproximately true) representations of a mind-independent world (realism).

In sum, there are a variety of different issues involved and positions taken regarding the nature of scientific laws and theories and, thus, a variety of different views of scientific methodology.

#7: The Aims and Goals of Scientific Ideas

Scientists and philosophers have surfaced several different aims or goals that are the explicit or implicit ends scientists intend to reach when they formulate theories. These goals are the values scientists and philosophers of science attach to theories that make them "good" theories. So understood, these goals are the epistemic values that define good scientific practice and constitute the cognitive aims scientists seek, or at least (assertedly) ought to seek. Sometimes different aims are compatible with each other, but sometimes they conflict. Here are some of the cognitive aims of science:

Truth or approximate truth. Realist scientists seek theories that are true or approximately true descriptions of the world. More will be said about this in chapter 4, but for now it should be pointed out that truth here means some sort of correspondence with the theory-independent world—the world is what makes our theories true or approximately true. The remaining goals are usually taken by scientific realists to be indicators that a theory is true or approximately true. For anti-realists, the following goals are thought to substitute for truth or approximate truth as the cognitive aims of science.

Theoretical simplicity. Theories often are sought that are theoretically simple. Given two rival theories that are equal in all other respects, the simpler theory usually is preferred. Two main problems are involved with the notion of simplicity. First, why should we believe that the world is such that it will be best described by simple theories? Simplicity is a notion that has its roots in medieval theism, wherein the world was viewed as a text from God. In a text, the most straightforward and simplest interpretation is accepted and an author is seen as an efficient mind who adopts the simplest means to communicate. Similarly, the world was understood as being metaphysically simple.

But apart from theism, simplicity is not easy to justify. (Even within theism, it is not clear that God will always do something in the simplest way, say by using the same design for all flying animals, because he may have other reasons for variety, like play and creativity.) To be sure, simple theories often have been better than more complex ones. Copernicus's theories were simpler than Ptolemy's. But this is not always the case. The ideal gas equation ($PV = nRT$) is simpler than the

Van Der Waals equation $[(P+a/V^2)(V-b) = nRT]$, but our growth in knowledge has favored the latter. Similarly, our growth in astronomy has favored more complex elliptical curves for planets over the relatively simple circular curved orbitals, the spherical shape theory of the earth has been replaced by a theory involving a more complicated shape, and so forth.

Second, it is sometimes hard to know what simplicity itself means. Theory A could be more simple than B if A has fewer basic assumptions, fewer basic kinds of entities postulated, fewer total entities regardless of the number of different kinds of entities, fewer independent variables, a smaller power of ten in any of the variables in the equations of the theory, or the smoothest curve when the theoretical laws are plotted on a graph. Regarding this last criterion, it sometimes happens that A is smoother if plotted on a normal Cartesian graph but B is smoother if plotted in polar coordinates. Which criterion shows theoretical simplicity? It seems that this must be decided on a case-by-case basis, and an element of arbitrariness is often hard to eliminate.

Empirical accuracy. A good theory should be empirically accurate, that is, its predictions or retrodictions ought to square with the data of observation and experiment.

The following issues arise regarding empirical accuracy: How reliable are our instruments in this field of science and what type of accuracy is to be expected empirically? Some areas of science are more qualitative, others more quantitative. Is the sheer quantity of data the important factor, or is the quality of special significance of the data important too? Sometimes theory A has been tested more or for some other reason gives more data than B, but B squares with some special range of data that are central to B (that is, B must square with these particular data because of their special significance to B). Also relevant is the presence or absence of a rival hypothesis that may or may not adequately describe the data special to the theory with which it is competing. Newton's laws of space, time, and motion had much more data on their side than did Einstein's model, especially when the latter was first introduced, but certain areas of data (e.g., Einstein's gravitational theory holds that light is attracted and bent by heavy bodies, and thus light from a distant fixed star whose apparent position was close to the sun would make the star appear as if it had moved a little away from the sun, a prediction that was observed by A. Eddington in 1919) were easy to square with Einstein's model but not with Newton's model.

Predictive success. Some argue that a good theory should predict data, although others hold that there is no significant difference between prediction and retrodiction or mere explanation of data; that is, the time factor is irrelevant as to whether the theory explains the data. For those who value prediction, difference can arise as to whether it is the sheer number of predictions, the variety of different kinds of things predicted, or the novelty of the prediction made by the theory being evaluated that matters.

Internal consistency and clarity. Good theories should be internally consistent, that is, they ought not to contain logical contradictions as judged by normal, two-valued logic (propositions are either true or false). Some scientists abandon this criterion in some areas of science (e.g., in quantum physics, some hold instead that the quantum world may obey a poly-valued logic—true, false, indeterminate). Further, good theories should have theoretical concepts that are clear and not vague or fuzzy in what they assert. Some would hold that clarity is achieved by quantifying the concept in question as far as that is possible (e.g., by defining "depression" as the state that falls within such-and-such a range in some psychological test); others would not see this as necessary. Further, some would hold that clarity is achieved by explaining some phenomenon in terms of a model that makes reference to the familiar, everyday world of macroentities (e.g., likening subatomic particles to tiny, spherical billiard balls). Others would argue that no special place should be given to the familiar as a criterion for clarity.

Adequacy in handling external conceptual problems. We have already discussed external conceptual problems. Suffice it to say that there are differences of opinion about the role of external conceptual problems in science and what disciplines these problems can come from.

Scope of reference. Many scientists hold that theory A is to be preferred to theory B if A has a broader scope of reference than does B. Thus, Newton's laws are to be preferred to Kepler's laws of planetary motion or Galileo's law of free fall because Newton's laws apply to a wider range of phenomena. It should be kept in mind that a branch of science sometimes must have a relatively clearly defined domain before scope considerations can be appropriate. That is, some criteria must exist for uniting a group of phenomena before it can be judged relevant or irrelevant that theory A extends into area X whereas theory B does not. For example, if terrestrial motion were thought to be essentially different from celestial motion, then Galileo's law of free fall could not be faulted for not having reference to motion in the heavens. On the other hand, often the very presence of a theory that does have broader scope not only makes such a theory preferable to a more narrow theory but also changes the way the domains in a given area are classified. Newton's laws changed the twofold classification of motion (celestial and terrestrial) and united them into one domain covered by the laws.

Fruitfulness in guiding new research. Sometimes a theory is judged to be preferable to a rival if it offers direction for the scientific community to explore new areas of research not indicated by the other theory. For example, the particle theory of light suggested that other areas of radiation phenomena formerly associated with waves (e.g., electrical phenomena) should be studied as particles. This in turn opened up new types of experiments and new kinds of explanations for electrical effects in terms of electrical particles called electrons.

Scientific practice or theories are formed, tested, and used according to accepted criteria for what science itself ought to be. Sometimes the scientific community has explicit or implicit rules governing its methods and definitions of what constitutes appropriate scientific practice. Methodological rules guide scientific practice (e.g., "Prefer double-blind to single-blind experiments," "Prefer experiments where instrument x was calibrated with procedure y"). Similarly, scientists can have views about what types of investigation or explanation are allowable (e.g., *in vivo* versus *in vitro* studies, contact-action versus field interpretations of forces, deterministic versus probabilistic explanations, naturalistic versus theistic modes of explanation, efficient versus final causes). Creation science is rejected by many scientists not only or even primarily because of data that are allegedly incompatible with it (most scientists probably do not even know enough about the details of creation science to form such a judgment), but primarily because it is thought to be a practice out of step with proper scientific methodology or modes of explanation.

In sum, four observations about the goals or aims of science surface from our list. First, when different goals are embraced by different scientists, then to that extent the scientists practice different scientific methodologies (e.g., scientist A does laboratory observations to get approximately true theories of the world; scientist B does them only to achieve simple, useful theories). Second, debates exist about the relative merits of the nine factors listed, and it sometimes happens that one of two rival theories will be better according to one criterion and the other will be better according to another. Third, within each individual criterion there are areas of debate and conflict (e.g., what is the proper view of simplicity?). Fourth, all the criteria listed involve philosophical considerations, especially the ninth. Thus, they illustrate the inseparability of science from philosophy (and in some cases, theology, e.g., criteria 2, 6, and 9).

Theology and Scientific Methodology

Throughout chapters 1 and 2, we have had occasions to observe that no sharp line can be drawn between science and various other fields of study. In particular, theology and science often overlap in the ways they form, use, and test hypotheses. In recent years several scholars have offered clear, powerful arguments that show the parallels between scientific and theological methodology.[34] In this section we will look briefly at one of these, offered by Patrick Sherry.[35]

Sherry argues that 1) there are significant similarities between religious explanations of sanctification and scientific explanations, 2) the phenomenon of sanctification and Christian transformation confirms Christian theism much as observational or experimental evidence confirms a scientific hypothesis.

In science, a theoretical term like *electron* has a certain meaning and refers to certain entities in the world with certain properties (negatively charged particles). The term *electron* functions in this way because it is tied to several other postulates in the entire theory of electricity, atomic phenomena, and the like.

Similarly, the concept of religious transformation gets its meaning by being tied to various ideas within a religious conceptual framework (e.g., the Holy Spirit produces transformation, certain spiritual practices facilitate this sanctification, such sanctification follows certain general laws governing its progression, and sanctification is a pledge or first fruit of what the future life will be like and, thus, can be used to predict short-range and long-range experiences). In short, certain religious phenomena (spelled out in the New Testament and in the lives of clear exemplars of religious transformation, e.g., St. Francis of Assisi) are effects brought about by the Spirit of God, and such effects provide evidence of God's activity and nature. God is the postulated cause of the radical effects, namely, spiritually powerful, morally virtuous people. Further, just as a scientific conceptual framework

34. John Warwick Montgomery, "The Theologian's Craft," in *The Suicide of Christian Theology,* by John Warwick Montgomery (Minneapolis: Bethany Fellowship, 1970), pp. 267–313; Richard Swinburne, *The Existence of God* (Oxford: Clarendon, 1979); George Schlesinger, *Religion and Scientific Method* (Dordrecht, Holland: D. Reidel, 1977), and *Metaphysics: Methods and Problems* (Totowa, N.J.: Barnes and Noble, 1983), pp. 48–96; William Wainwright, *Philosophy of Religion* (Belmont, Calif.: Wadsworth, 1988), pp. 48–62; John Polkinghorne, *One World: The Interaction of Science and Theology* (Princeton, N.J.: Princeton University Press, 1986), pp. 6–42.

35. Patrick Sherry, *Spirit, Saints, and Immortality* (Albany: State University of New York Press, 1984), pp. 31–63.

will tie together past, present, and future phenomena, so does the Christian framework. The concept of "sanctification" ties together the past historical facts about Jesus Christ, the present activities of his Spirit in religious transformation, and the future state of life after death. All three receive mutual support from each other and admit of a degree of independent verification when considered on their own. Thus, the religious conceptual framework relates various phenomena to each other and to various core facts in the past, present, and future, and seeks to assign a cause adequate to explain the relevant data, which, in turn, serve as confirmation for that hypothetical cause.

This religious conceptual framework admits of general predictions and can, in principle, be falsified. Sherry points out that Christian theism is not equally compatible with all possible worlds. If the Christian faith never produced powerful examples of religious transformation when certain spiritual laws were properly put to the test, then this would count significantly against the truth of Christian theism and some of its specific doctrines associated with divine grace, prayer, and so forth.

In sum, religious believers point to something visible and recognizable to everyone (the presence of saintliness in certain individuals who have seriously practiced the spiritual laws associated with the Christian conceptual framework), they use special terms (e.g., *grace, sanctification*) embedded in their ontological model of the world to describe and explain saintliness, and the presence of spiritual saints and religious transformation provides evidence for the truth of that framework (as its absence would tend to falsify it).

Sherry does not claim that his argument for Christian theism is capable of offering mathematical precision in its predictions and law-like relations. But neither does every area of science, especially those broad scientific conceptual schemes (e.g., physicalism) that approach a world view in scope. But Sherry has shown that religious conceptual schemes use some of the same logical structures in the formation of their hypotheses, the postulation of causes, the use of those hypotheses to explain data, and the role of those data in confirming those hypotheses.

Summary and Conclusion

The main purpose of this chapter was to consider the claim that there is a fairly clear thing called the scientific method that distinguishes science from other fields of study like theology. We have seen that this claim is naive because it fails to take seriously both the nature of human action in general and the various issues involved in analyzing scientific methodology. To be more specific, three conclusions have been reached.

First, the question of whether there is something called the scientific method is itself a philosophical question (and a historical one), not primarily a scientific question. Science enters the point of providing examples of scientific practice to analyze. But the analysis of scientific methodology is properly the domain of historians and philosophers of science.

Second, there is no single thing called the scientific method. Such a notion is a myth for at least three reasons: 1) There is a debate about the relative merits of inductivism and a more eclectic model of science. 2) There are different areas of debate within the eclectic model itself. 3) Different areas of science and different stages in the development of some particular areas of science use different aspects of the various areas we have discussed in analyzing the eclectic mode.

Third, it is more proper to say that there is a family of methodologies used by various areas of science, methodologies that are only capable of broad characterization. So understood, disciplines outside science like theology can be shown to use various aspects of scientific methodology. The idea that science is a rational, truth-seeking discipline and theology is not is a widespread cultural myth. This myth often is promulgated by contending that science gains its status by its privileged use of a specific methodology not available to theology. But such a claim is itself a myth—the myth of ostrich scientism—that needs to be laid to rest.

The Limits of Science 3

> The theorist who maintains that science is the be-all and end-all—that what is not in science textbooks is not worth knowing—is an ideologist with a peculiar and distorted doctrine of his own. For him, science is no longer a sector of the cognitive enterprise but an all-inclusive world-view. This is the doctrine not of *science* but of *scientism*. To take this stance is not to celebrate science but to distort it. . . .   —Nicholas Rescher, *The Limits of Science*

> A successful argument for science being the paradigm of rationality must be based on the demonstration that the presuppositions of science are preferable to other presuppositions. That demonstration requires showing that science, relying on these presuppositions, is better at solving some problems and achieving some ideals than its competitors. But showing that cannot be the task of science. It is, in fact, one task of philosophy. Thus the enterprise of justifying the presuppositions of science by showing that with their help science is the best way of solving certain problems and achieving some ideals is a necessary precondition of the justification of science. Hence philosophy, and not science, is a stronger candidate for being the paradigm of rationality.
> —John Kekes, *The Nature of Philosophy*

One evening at a dinner party I was introduced to a man who had been an engineer for twenty-five years and was finishing his doctorate in physics. When he learned that I was a philosopher and theologian, he engaged me in a discussion of the Alice-in-Wonderland character of both of my disciplines. According to him, only science is rational; only science achieves truth. Everything else is mere belief and opinion. He went on to say that if something cannot be quantified or tested by the scientific method (he never told me what that was), it cannot be true or rational.

Unfortunately, this opinion is widely shared in Western culture. It goes by the name of *scientism* or, as John Kekes calls it, *scientific imperialism.* According to this view, science is the very paradigm of truth and rationality. If something does not square with currently well-established scientific beliefs, if it is not within the domain of entities appropriate for scientific investigation, or if it is not amenable to scientific methodology, then it is not true or rational. Everything outside of science is a matter of mere belief and subjective opinion, of which rational assessment is impossible. Science, exclusively and ideally, is our model of intellectual excellence.[1]

But however widely this opinion may be held, it is nonetheless a cultural myth that is patently false. The major burden of this chapter is to show that there are various limits to science that, taken together, provide a powerful case against scientism.

What do we mean by a *limit to science?* To begin with, we do not refer to three kinds of limits to science, as interesting as they may be in other contexts:

First, we will not discuss practical limits to science. Practical limits to science enter into debates about how far science might go. For example, scientific progress involves a process of technological escalation requiring increasingly elaborate technology, which in turn requires greater and greater economic and social effort directed toward science. Thus, scientific growth might be practically limited by finite social, economic, and environmental resources.

Second, we will not discuss the cultural-moral limits to science. Limits of this kind arise in debates about conflicting goals that our culture ought to have, debates in which science is blamed for ". . . causing a deterioration in the quality of life and dashing our hopes of its improving, of despoiling the environment and the like."[2] It is sometimes said, for example, that the space program is an immoral diversion of national funds, since other goals are more easily justifiable from a moral point of view (e.g., eliminating poverty).

Third, we will not discuss in detail the limits one science places on another. For example, the various physical components of living organisms and their chemical interactions are appropriately classified in the domain of chemistry. But the fact that some given chemical ensemble exists for some given organism, for example, the question of why a jellyfish has its particular chemical makeup and structure, is a question not primarily for chemistry but for biology. The biologist

1. This attitude is partly responsible for the demise of the humanities in the university and the conceptual relativism, what E. D. Hirsh calls "Kantianism gone mad," in those disciplines. See Carl F. H. Henry, *The Christian Mindset in a Secular Society* (Portland, Ore.: Multnomah, 1984), pp. 81–96.

2. Peter B. Medawar, *The Limits of Science* (New York: Harper and Row, 1984), p. 12.

answers the question by focusing on the history of the development of the organism, its role in an environmental ecosystem, and so forth. But these are distinctively biological questions using biological terms, laws, and explanations. Thus the presence of such issues regarding, say, a jellyfish, currently places a limit on how far chemistry can go in explaining all we want to know scientifically about jellyfish. This is an example of one science limiting another and, though it is interesting, we will not investigate in detail limitations of this type.

A limit to science as we will be using it is, roughly, an explanation or answer to some problem that properly lies outside the boundaries of scientific explanation and is cognitive in nature, that is, it is in principle a rational issue whose solution can be true or approximately true. I do not mean for this to be an airtight definition of a limit to science, since such definitions are, as we have seen, difficult to secure and no clearcut boundary exists between science and nonscience. Nevertheless, this is an adequate characterization for our purposes. So understood, limits of science can be classified in the following way:[3]

Presuppositions that underlie science. These are philosophical preconditions on which science depends for rational justification.

Cognitive issues outside of science. These cognitive issues do not constitute preconditions for science but lie outside the domain of scientific investigation, and, generally speaking, come in two types: *extraneous cognitive issues*, whose solutions are irrelevant and do not clearly interface with science, and *external conceptual problems*, whose solutions lie primarily outside of science but nevertheless interface with science such that scientific ideas may conflict with them and require modification in light of them. These problems arise primarily, though not exclusively, when science is understood as scientism.

Limits within science. These issues are not preconditions to science and do not arise outside of science, but arise from science itself or from some other field like philosophy and limit the very practice of science itself.

Our analysis of the limits to science rests on two assumptions. First, we assume that we understand science in terms of scientific realism. For our purposes this means we see science as a rational, truth-seeking discipline, that is, one purporting to give, in a rationally justifiable way, a true, or approximately true, description of the way the world really is.

3. For alternative classifications of the limits of science, see ibid., p. 67; John Kekes, *The Nature of Philosophy* (Totowa, N.J.: Rowman and Littlefield, 1980), pp. 147–63; Nicholas Rescher, *The Limits of Science* (Berkeley: University of California Press, 1984), pp. 1–4; Del Ratzsch, *Philosophy of Science* (Downers Grove: Inter-Varsity, 1986), pp. 97–105.

Second, we assume that we have a fairly clear idea of what science is and that this idea will not change appreciably in the future. We have already seen that there is no clear boundary between science and nonscience, but a continuum, and that scientific theories of the world, as well as our conceptions of what we take science to be, can change with time. But we can still identify cases at one end of the continuum that are clear practices of science, and we can reasonably assume that scientific practice in the future will form, test, use, and interpret theories in a manner similar to our views today, especially given our assumption of scientific realism.[4] Before we turn to an analysis of various presuppositions of science, let us look first at an important limit to science—the definition of science itself.

The Definition of Science and Self-Refutation

A statement is about its subject matter. The statement "Grandview is a city in Missouri" is a statement about the city Grandview, "electrons have negative charge" is about electrons, "education is an interesting field of study" is about the discipline of education. Some statements refer to themselves; that is, they are included as an item in their own field of reference. "All sentences of English are short" makes reference to all English sentences including itself. When a statement includes itself within its field of reference and fails to satisfy itself (i.e., to conform to its own criteria of acceptability), it is self-refuting.[5] For example, "There are no true statements," "I do not ex-

4. Rescher argues that future science is so unpredictable that we have no idea whatever as to what the future questions of science will be and, thus, we can set no limits to the results of science. See Rescher, *Limits of Science*, pp. 95–132. However, Rescher is an antirealist who sees the aims of science to be prediction and control of nature, not a progressive convergence toward a true picture of the world. I am assuming scientific realism, which allows for future science to differ partially from present "mature" cases of science but also implies that future science will resemble the practices and beliefs of present mature sciences. Furthermore, Rescher runs the risk of making "science" so plastic as to have no identifiable, stable meaning beyond the mere formal stipulation that science is the field that seeks to control the world and predict phenomena.

5. For good treatments of self-refutation, see Michael Stack, "Self-Refuting Arguments," *Metaphilosophy* 14 (July/October 1983): 327–35; George Mavrodes, "Self-Referential Incoherence," *American Philosophical Quarterly* 22 (January 1985): 65–72; Carl Kordig, "Self-Reference and Philosophy," *American Philosophical Quarterly* 20 (April 1983): 207–16; Joseph Boyle, "Self-Referential Inconsistency, Inevitable Falsity, and Metaphysical Argumentation," *Metaphilosophy* 3 (January 1972): 25–42; Joseph Boyle, Germain Grisez, and Olaf Tollefsen, *Free Choice: A Self-Referential Argument* (Notre Dame: University of Notre Dame Press, 1976), pp. 122–152.

ist," and "No one can utter a word in English" (uttered in English) are all self-refuting. If they are false, they are false. If they are assumed to be true, what they assert proves them false, and since nothing can be both true and false at the same time, they turn out to be false. Either way, they are necessarily false. The facts that falsify them are unavoidably given in the statements themselves. They cannot be true.

A dogmatic claim of scientism (e.g., "only what can be known by science or quantified and tested empirically is true and rational") is self-refuting. The statement itself is not a statement of science, but a second-order philosophical statement about science. The statement cannot be tested empirically, quantified, and so on. Another way to put this is to say that the definition, aims, and justification of science are philosophical presuppositions *about* science and cannot be validated *by* science. One cannot ask science to justify itself any more than one can pull oneself up by one's own bootstraps. Justifying science by science is question begging. The validation of science is a philosophical issue, at least in part, and any claim to the contrary will itself be philosophical.

We have already seen in chapters 1 and 2 that defining science is a philosophical issue. Philosophy is, among other things, a second-order discipline about first-order disciplines, including science. Scientists investigate questions about DNA, covalent bonds, ecosystems, the nature of water, and so forth. Philosophers investigate questions about science like those in the first two chapters: "What are the necessary and sufficient conditions for something to count as science?" "What procedures should a scientist follow in investigating nature?" "What counts as confirmation of a theory?" "How do theories explain?" "Should we be scientific realists or antirealists?" To ask such questions is to take a vantage point outside science, not within science. This is a philosophical vantage point, even if the one asking the questions is herself a scientist. When she asks these questions she does so as a philosopher, not as a scientist. There is a difference between doing science and thinking about what science is and how it ought to be done. The latter questions are philosophical.[6]

This approach to defining science is called an external philosophy of science. With science itself the datum, one applies a general philosophical understanding of reality, knowing, logical structure, and so on, to episodes of science, evaluating the episodes as good or bad sci-

---

6. For more examples of the difference between doing science and philosophical thinking about science, see John Losee, *An Historical Introduction to the Philosophy of Science*, 2d ed. (Oxford: Oxford University Press, 1980), pp. 1–4; Ernan McMullin, "Alternative Approaches to the Philosophy of Science," in *Scientific Knowledge*, ed. Janet A. Kourany (Belmont, Calif.: Wadsworth, 1987), pp. 3–19.

ence with justified or unjustified theories. The historian of science is also involved in defining science, for he provides detailed episodes for the philosopher to evaluate. Both the philosophy of science and the history of science are second-order inquiries about a first-order range of data—science itself.[7]

Recently, thinkers like Wilfred Sellars and W. V. O. Quine have advocated an internal philosophy of science.[8] The details of this approach are beyond the scope of our investigation. Briefly, this understanding of philosophy and science attempts to subsume the former under the latter. Philosophy is just a branch of science. Science is its own justification.

But this viewpoint is inadequate. It begs the question by assuming that science is self-justifying, and it removes the distinction between descriptive questions and normative questions. Scientists, say sociologists, could describe scientific practice, but philosophy asks normative epistemological questions about science. More will be said about this approach to science and philosophy later when we discuss conceptual problems with evolutionary theory. For now, we restate our main point: the problem of defining and justifying science is a normative, second-order philosophical (and historical) problem that takes science itself as the first-order datum. Let us turn to an investigation of the various presuppositions of science.

Presuppositions of Science

If P is a presupposition of Q, then P is fundamental for Q, that is, P is a necessary condition for Q. If one abandons P, then he must abandon Q. Several presuppositions or preconditions must be assumed if science, as it is understood by scientific realists, is to be possible. As John Kekes has observed:

7. For more on the relationship among history of science, philosophy of science, and science itself, see John Losee, *Philosophy of Science and Historical Enquiry* (Oxford: Oxford University Press, 1987).

8. Quine is the main advocate of this position. See W. V. O. Quine, *Ontological Relativity and Other Essays* (New York: Columbia University Press, 1969), pp. 69–90. For overviews of Quine's thought that include criticisms, see Jonathan Dancy, *Introduction to Contemporary Epistemology* (New York: Basil Blackwell, 1985), pp. 92–95, 233–41; Christopher Hookway, *Quine* (Stanford, Calif.: Stanford University Press, 1988), especially chapters 4, 11, and 12; Hilary Kornblith, ed., *Naturalizing Epistemology* (Cambridge, Mass.: MIT Press, 1985). See also Robert Audi, "Realism, Rationality, and Philosophical Method," *Proceedings and Addresses of the American Philosophical Association*, supplement to vol. 61, no. 1 (September 1987): 65–74.

Science is committed to several presuppositions: that nature exists, that it has discoverable order, that it is uniform, are existential presuppositions of science; the distinctions between space and time, cause and effect, the observer and the observed, real and apparent, orderly and chaotic, are classificatory presuppositions; while intersubjective testability, quantifiability, the public availability of data, are methodological presuppositions; some axiological presuppositions are the honest reporting of results, the worthwhileness of getting the facts right, and scrupulousness in avoiding observational or experimental error. If any one of these presuppositions were abandoned, science, as we know it, could not be done. Yet the acceptance of the presuppositions cannot be a matter of course, for each has been challenged and alternatives are readily available.[9]

Let us look at some of the presuppositions of science.

The Existence of the External World

Science assumes that there is a distinction between the observer and the observed and that there really does exist an external, mind-independent (or language-independent) world wherein particular objects (atoms, organisms, planets) have real, independent existence and stand in various types of relationships with other objects. This assumption is a fairly common-sense approach to the world. Nevertheless, it has been challenged.

First, the history of religion and philosophy has included a number of views, like pantheism and absolute idealism, that deny the full reality of an external, mind-independent, material universe of independently existing things. Our purpose here is not to investigate the relative merits of pantheist or absolute idealist arguments, though some of them are quite sophisticated; rather, it is to point out that one must engage in philosophical argumentation to justify this precondition of science in light of these alternative pictures of the world.

Second, from the time of René Descartes (1596–1650) through the British empiricists (especially Berkeley and Hume) to the empirically oriented logical positivists of this century, another objection has been raised against the existence of the external, material world. This objection comes in two steps. First, it is assumed that between an observing subject and the object of observation stands an intermediate object that is the direct object of our observational experiences. This intermediate object has been variously named—a sense datum, a sense image, or a simple idea (like a patch of red). According to this theory

9. Kekes, *Nature of Philosophy*, pp. 156–57.

of perception, objects in the world (if they exist at all), like trees and lions, are never directly observed by anyone. Instead, what we directly see are our sensory images or mental pictures of trees and lions. Sense images are intermediate objects intervening between us and the world, and we directly perceive only these.

Second, it follows that I can never get outside my own sensory experiences to the external world itself. I can never compare my ideas with the world, I can never tell what causes my sense images, and so forth.[10] For some thinkers, like Berkeley, this state of affairs has the following consequence. It becomes cognitively meaningless to use the word *exists* for something outside the pale of experience itself. To say "x exists" is merely to assert that if I orient my eyes in certain ways, I will have a potentially infinite set of sensory experiences. If x = a red chair, then when I assert that x exists I merely affirm that if I walk around in a room, I can have a series of red-chair-type experiences. It makes no sense, on this view, to ask the further question, "What is it that exists and that is causing these sense experiences?" for this question assumes that "exists" has some further meaning (e.g., "The chair exists in the external world and causes my sensory experiences of it.") besides the statement of my actual or possible sense experiences (e.g., "the chair exists" means simply, "I am having a red-chair-type experience now, and I could have another red-chair-type experience larger than this one if I moved over there," and nothing more).

On this view, the use of *exists* in "the external world exists" is meaningless. Again, our purpose is not to argue for or against this viewpoint. I merely note that it has attracted no small number of followers, and while I am not among them, the arguments for and against it are philosophical in nature and presuppositional to science.

Finally, it should be noted that some contemporary physicists (though far from the majority) assert that quantum phenomena indicate that consciousness or the process of observation itself creates the observer's own reality. On this view, the world becomes "real" only when and while it is consciously observed. The notion of a mind-independent, materially existing external world is bogus.[11] This viewpoint is not accepted by all, or perhaps many, scientists and philosophers, but its presence calls into question this presupposition of modern science, and the debate about it is philosophical.

10. For the best recent defense of this view of perception, see Frank Jackson, *Perception: A Representative Theory* (Cambridge: Cambridge University Press, 1977). A nontechnical introduction to various theories of perception can be found in Robert Audi, *Belief, Justification, and Knowledge* (Belmont, Calif.: Wadsworth, 1988), pp. 8–29.

11. Cf. Nick Herbert, *Quantum Reality* (Garden City, N.Y.: Doubleday, 1985), pp. 1–29.

*The Orderly Nature of the External World
and Its Knowability*

Another presupposition of science is that the world exhibits various kinds of order that can be known.[12] In other words, the world, or some range of phenomena within it, has a mind-independent order that we sometimes discover. For example, scientists sometimes investigate what Dudley Shapere calls ordered domains—groups of phenomena in which types of items are classified and those classes are themselves arranged in some pattern, perhaps a series, according to some principle or property.[13] The periodic table is an example. Chemical elements are arranged according to their atomic number and fall into natural, repeating groups (e.g., the inert gases or transition metals) based on sequentially ordered properties of their bonding tendencies.

However, the orderly nature of the world as it is in itself has been challenged, chiefly by Immanuel Kant (1724–1804) and his followers, including Harvard philosopher Hilary Putnam.[14] On this view, there is a noumenal world that exists in itself, but it is not the world of our experience and we can know nothing about it as it is in itself. The world that we experience, the phenomenal world, is a mixture of the input it receives from the noumenal world (whatever that is) and the order imposed on that input by the senses and mind of the knowing subject. The noumenal world is an undifferentiated given, forever outside our gaze. The order in the world of our experience—the distinction between a cause and an effect, the unity of a single thing and the plurality of different things given to us in experience, the distinction between an object like an apple and the attributes of the object like

12. Rescher argues that future science may be so unlike present science that the assumption of an orderly world that is uniform may not be part of science. See Rescher, *Limits of Science*, pp. 104–9. See note 4 for a brief criticism of Rescher. Even when science denies that some aspects of the world are orderly, say, certain quantum phenomena, it assumes that higher-order regularities exist that can be known, e.g., that the laws of probability and the equations of quantum physics apply to reality. So even if future science denies some area of order we now accept, such a denial will presuppose another area of knowable order and regularity. For more on the intelligibility of the world as an assumption for science, see George Schlesinger, *The Intelligibility of Nature* (Aberdeen, Scotland: Aberdeen University Press, 1985).

13. See Dudley Shapere, "Scientific Theories and Their Domains," in *The Structure of Scientific Theories*, ed. Frederick Suppe, 2d ed. (Urbana, Ill.: University of Illinois Press, 1977), pp. 518–70.

14. For an overview of this approach to science, see Kurt Hubner, *Critique of Scientific Reason*, trans. Paul R. Dixon and Hollis M. Dixon (Chicago: University of Chicago Press, 1983), pp. 3–12; Hilary Putnam, *Philosophical Papers, Vol. 3: Reason, Truth, and History* (Cambridge: Cambridge University Press, 1981), pp. 49–74, especially pp. 60–64.

redness—is all contributed by the knowing subject, who imposes order on the noumenal world to synthesize the world of experience.

On this view, the task of science is to discover the general forms or structures of our sensory experience (the phenomenal world), structures that are only apparent to us, that is, structures we impose on the noumenal world to order the phenomena of our experience, not structures we discover in noumenal reality itself.[15] The laws of science that state the way different phenomena relate to one another do not describe orderly relations that we discover in the mind-independent world. Rather, the laws are part of the structure of our consciousness of the phenomenal world, they are the way we must experience the phenomenal world, the way in which the categories of the knowing subject take the chaotic input from the noumenal world, subsume and classify it, and create an ordered world of experience (the phenomenal world).

If Kant and his contemporary followers are correct, then science, as it is understood in classical realist terms, is not possible because one of the major preconditions for science, the existence of order that can be discovered in the (mind-independent) world (and not created by the knowing subject itself) is an illusion.

The Uniformity of Nature and Induction

Science engages in inductive, ampliative inferences. In an ampliative inference, the conclusion fills out and adds to the premises. In a nonampliative inference, the conclusion is already contained, implicitly or explicitly, in the premises and thus adds nothing to them. "All men are mortal, Socrates is a man, therefore Socrates is a mortal" is an example of a nonampliative inference.

Inductive inferences are ampliative. For example, "All observed ravens are black, therefore all ravens are black" is an inductive inference. The conclusion contains more information than does the premise since it refers to all ravens, while the premise refers only to observed ravens. Scientists make inductive (ampliative) inferences all the time. From the fact that all observed copper rods expand when heated, or all observed increases in pressure cause increases in the temperature of a gas at constant volume, we conclude that all copper rods expand when heated and all increases in pressure cause an increase in temperature at constant volume.

15. Cf. Suppe, *Structure of Scientific Theories*, pp. 6–15; Max Delbruck, *Mind from Matter: An Essay on Evolutionary Epistemology*, ed. Gunther Stent and David Presti (Palo Alto: Blackwell Scientific Publications, 1986), pp. 7–8, 117–18.

Inductive inferences reason from past cases of a phenomenon to future cases, from observed cases to unobserved cases. But how do we justify such ampliative, inductive inferences? This is the problem of induction, and it was first raised by David Hume. Here is Hume's statement of it:

> It is impossible, therefore, that any arguments from experience can prove this resemblance of the past to the future, since all these arguments are founded on the supposition of that resemblance. Let the course of things be allowed hitherto ever so regular, that alone, without some new argument or inference, proves not that for the future it will continue so. In vain do you pretend to have learned the nature of bodies from your past experience. Their secret nature, and consequently all their effects and influence, may change without any change in their sensible qualities. This happens sometimes, and with regard to some objects. Why may it not happen always, and with regard to all objects? What logic, what process of argument secures you against this supposition? My practice, you say, refutes my doubts. But you mistake the purport of my question. As an agent, I am quite satisfied in the point; but as a philosopher who has some share of curiosity, I will not say skepticism, I want to learn the foundation of this inference.[16]

Hume's point is to ask how it is ever possible to justify any inductive inference. Another way to state the problem of induction can be seen in figure 3.1. The symbol t stands for a time marking the present moment. Until now we have been living in world W_1 where ravens are black, copper expands when heated, and so on. The chart contains three possible worlds that are the same up until the present moment but diverge in the future: W_1–W_3, W_1–W_2, and W_1–W_4. The first of these worlds (W_1–W_3) is one wherein the future is uniform with the past and inductive inferences are justified—copper continues to expand when heated and ravens continue to be black. But the other two worlds are such that the future takes different, unpredictable courses such that inductive inferences are not justifiable. The problem of induction now can be stated in this way: How do we know that we are in W_1–W_3, and not W_1–W_2 or W_1–W_4?

There have been several attempts to solve the problem of induction. Let us look briefly at three of them.

16. David Hume, *An Inquiry Concerning Human Understanding* (1748; Indianapolis: Bobbs-Merrill, 1965), pp. 51–52 (section 4.2 in the original). For a contemporary analysis of Hume's problem of induction, see Wesley C. Salmon, *The Foundations of Scientific Inference* (Pittsburgh, Pa.: University of Pittsburgh Press, 1967).

Figure 3.1
The Problem of Induction

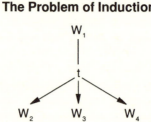

THE INDUCTIVE JUSTIFICATION OF INDUCTION

Some have attempted to justify inductive inferences by saying that such inferences can themselves be justified inductively—induction has worked well in the past and we can conclude, therefore, that induction will continue to work in the future.

Obviously, this solution is inadequate. First, it is clearly question begging and viciously circular, for it assumes induction to prove induction. The problem of induction can be rephrased with regard to inductive inferences themselves: How does the fact that past examples of induction have worked well justify our belief that such inferences will continue to work in the future?

Second, the inductive justification of induction is an example of a self-validating procedure, and such a procedure is usually suspect, to say the least. Self-validating procedures can "validate" bad procedures. Here are some examples:

Example 1: A crystal ball that says "In the future I will be accurate."

Example 2: If affirming the consequent is valid, then grass is green. Grass is green.
Therefore, affirming the consequent is valid.

Example 3: Counterinduction—If all past As have been B, then probably the next A will not be B. A = Examples of counterinductive inference, B = false.

Example 1 is a case where a crystal ball predicts that it will be accurate in the future. But how do we know this? Because the crystal ball predicts it. How do we know that crystal ball predictions can be trusted? Because in the future the crystal ball will be accurate, just as it presently predicts. Example 2 is itself an example of affirming the consequent. In the first premise, "affirming the consequent" and "grass is green" are called the antecedent and the consequent, respectively, and it is a logical fallacy to affirm the consequent. But if one is allowed to do so, then one can validate even acts of affirming the consequent.

Example 3 is closer to our present concern. It is an example of a counterinductive rule, that is, a rule that is the opposite of induction. It says that if all past examples of counterinduction have been false, then the next example will be true. Why? Because of the counterinductive rule itself. The reasoning is circular: only by assuming that counterinductive argument is valid can we conclude that this counterinductive argument for the validity of counterinductive argument is itself valid.

These three examples illustrate the fact that a self-validating procedure can validate something illegitimate. So in general, such procedures are suspect; in particular, how can we know that the inductive justification of induction is not as vicious as one of our three examples?

THE ORDINARY-LANGUAGE JUSTIFICATION OF INDUCTION

Other philosophers have attempted to "justify" induction by claiming that inductive inferences are simply what we mean by using ordinary English words like *rational, evidence,* and so on. In English we use the word *evidence* simply to mean *inductive evidence*. The word *rational* means the following: "Rational $=_{def.}$ A belief is rational if and only if it is formed according to rule R where R is simply inductive inference (i.e., If all observed As are B, then probably the next A will be B)."

The main problem with the ordinary-language defense of induction is that rationality is a matter not of behaving according to some word in a language, but of approaching beliefs in a way that is intrinsically justifiable in a normative sense. In other words, what one wants is really to *be* rational, to have beliefs one *ought* to have, not merely to behave according to what English language users mean by the word *rational*. After all, English could have developed among irrational people.

Furthermore, the ordinary-language argument fails to answer the problem of induction for another reason. Suppose we introduce a new word into English as follows: "Brashional $=_{def.}$ A belief is brashional if and only if it is formed according to rule R' where R' is simply counterinductive inference (i.e., if all observed As have been B, then probably the next A will not have been B)."

Brashional is defined according to counterinduction, in which inductive inferences break down. Hume's problem of induction can now be restated as follows: Why should we prefer to be rational instead of brashional? The ordinary-language argument does not seem to be able to answer this question.

THE SYNTHETIC A PRIORI SOLUTION

A third solution, which I favor, is to adopt the uniformity of nature (or perhaps of classes of things like the class of electrons) as a synthetic a priori. In order to understand this, look at figure 3.2.

Statements can be analytic or synthetic. An analytic statement is one wherein the predicate adds nothing to the subject, but is contained in the subject or is a synonym for it. "Everything red and round is red" and "All bachelors are unmarried males" are analytic statements. The subject "red and round" contains the predicate "red" within it and the subject "bachelors" is by definition synonomous with "unmarried males." One does not verify analytic statements by looking at the world. No one would survey all the bachelors of Virginia to see if, in fact, they were all unmarried males. In fact, analytic statements can be true even if they make no claims about things that really exist in the world. "All unicorns are one-horned horses" is analytically true even though no unicorns exist.

By contrast, synthetic statements have predicates that do add something to the subject. "All ravens are black" is an example; "black" is not part of what it means to be a raven. One discovers that the statement is true by investigating ravens.

Statements can also be a priori or a posteriori. The former are known without appealing to sensory experience for their justification whereas the latter require such an appeal. "Red is a color," "2 + 2 = 4," and "bachelors are unmarried males" are examples of a priori statements; "all ravens are black" is an example of an a posteriori statement.

A synthetic a priori statement makes a claim about the world, is necessarily true, and can be known a priori. Examples of alleged synthetic a priori statements are: "Red is a color," "Something cannot be red and blue all over at the same time," and "Mercy is a virtue."

According to the synthetic a priori solution to the problem of induction, one claims to adopt the uniformity of nature (the future will resemble the past) as a necessary truth that one simply knows by immediate rational intuition. Or one can claim to know that all copper

Figure 3.2
Kinds of Statements

Analytic	Synthetic
A priori	A posteriori

rods will continue to expand when heated because he can see immediately that expanding when heated is a property that is essential to copper.

Philosophers who reject this approach deny either that properties exist and that things have natures or essences or that we have anything like a faculty for rational intuition.

As stated earlier, the point here is not to argue for a solution to the problem of induction but to illustrate the problem itself and the philosophical nature of its solution. Thus, the uniformity of nature and the justification of induction are philosophical presuppositions of science.[17]

The Laws of Logic, Epistemology, and Truth

Three closely related areas of philosophical study that involve presuppositions of science are logic, epistemology, and truth.

Scientists use the laws of logic to make inferences in forming, testing, and using theories. Logic is a branch of philosophy that studies, among other things, the various laws of logic that distinguish valid from invalid inferences. A valid deductive argument is one in which the truth of the premises guarantees the truth of the conclusion. If the premises of a valid deductive argument are true, then the conclusion must be true. An invalid deductive argument is one in which the conclusion does not necessarily follow from the truth of the premises. One of the laws of logic is *modus ponens*:

$$H \rightarrow Q$$
$$\frac{H}{Q}$$

H could be "a tornado is coming" and Q could be "a drop in air pressure will occur." If it is true that the coming of a tornado allows one to infer that a drop in air pressure will occur, and if it is true that a tornado is coming, then we can infer that a drop in air pressure will occur. In contrast, invalid inferences exhibit various logical fallacies. For example, from "All dogs are mammals" and "All cats are mammals" we cannot infer "All dogs are cats." Science does not study the laws of logic, philosophy does. Yet science uses those laws. Thus the laws of logic constitute philosophical presuppositions of science.

Issues in epistemology present yet another set of presuppositions to

17. For more on inductive reasoning in science, see Rom Harre *The Philosophies of Science: An Introductory Survey* (Oxford: Oxford University Press, 1972), pp. 35–61.

science.[18] Epistemology is a branch of philosophy that focuses on these questions: What is knowledge or justified belief? Can we have knowledge or justified belief (the question of skepticism)? What can we know or have a justified belief about (the question of the scope of knowledge)? How do we know or justify our beliefs (the question of criteria for knowledge)?

It should be clear that if knowledge in general is impossible, then so is scientific knowledge. Again if we cannot know anything beyond the immediate deliverances of the senses, then we cannot know our claims about unobserved or unobservable theoretical entities like electrons or magnetic fields. Epistemology is conceptually prior to science because questions about knowledge in general are normative, nondescriptive questions, and they are prior to questions about any given species of knowledge (e.g., scientific knowledge).

Finally, science assumes that truth is possible and that truth is a certain sort of thing, at least if we understand science along the lines of scientific realism. Most scientific realists assume that truth is to be understood in terms of some form of the correspondence theory. There are different versions of this theory, but the essence of it is that truth is a relation of correspondence that obtains between our views of the world and the world itself. Truth obtains when what we say about the world describes, in fact, the way the world really is.[19]

There are three main issues involved in spelling out the correspondence theory of truth (see fig. 3.3).

Figure 3.3
The Correspondence Theory of Truth: Three Issues

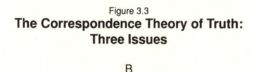

18. The relationship between epistemology and science involves a discussion of two schools of thought regarding the structure of justification of our beliefs—foundationalism and coherentism. The details of this discussion are beyond the scope of our investigation. For an overview of foundationalism and coherentism as epistemological strategies and for a guide to further literature on the subject, see J. P. Moreland, "Foundationalism and Coherentism: A Comparison," *Bulletin of the Evangelical Philosophical Society* 9 (1986): 21–30.

19. For overview discussions of the correspondence theory of truth and its rivals, see Keith Lehrer, *Knowledge* (Oxford: Clarendon, 1974), pp. 24–48; D. W. Hamlyn, *The Theory of Knowledge* (Garden City, N.Y.: Doubleday, Anchor Books, 1970), pp. 112–42; A. R. Lacey, *Modern Philosophy* (Boston: Routledge and Kegan Paul, 1982), pp. 27–47.

Issue A involves what we say it is that corresponds to the world when we have the truth. Three main candidates usually are appealed to.[20] The first is sentences—strings of physical markings on a sheet of paper (or, for audible sentences, sounds) in some language. The second candidate is statements. A statement is a speech act, that is, an act of speaking (or writing) where a person performs (or attempts to perform) an act of communicating some content like "the dog is in the yard." Speech acts are events. The third candidate is propositions—abstract entities that cannot be seen but exist and are the contents of sentences and statements and the subject matter of logic.

Issue B involves an analysis of the relation of correspondence itself. Just exactly what sort of a thing is correspondence? It certainly does not seem to be a physical relation like "larger than" or "to the left of." When my concept of a human being "corresponds to" what humanness really is, what is this relationship? Some reduce the relation of correspondence to that of identity. The concept of a human being itself consists of all and only those properties that human beings have. The concept of humanness is itself humanness. Others prefer to say that correspondence is an intentional relation of aboutness or ofness. That is, the concept of humanness is a mental thing. The concept of humanness is *about* humanness. The concept itself has none of the properties of humanness, but is about those properties.

Issue C involves analyzing what our propositions and concepts correspond to when we have the truth. Usually, philosophers say that our true propositions correspond to facts or states of affairs in the world (some identify facts with states of affairs and some distinguish them). For example, my thought that the apple is red is true if and only if it corresponds to the state of affairs constituted by the apple's being red.[21]

In sum, science (at least as viewed by scientific realists) assumes some sort of correspondence theory of truth, which in turn requires philosophical solutions to the various issues involved in unpacking what the correspondence theory is supposed to be. If this cannot be done, then some argue that the correspondence theory should be abandoned and, with it, scientific realism as a model of science. Furthermore, if the best solution to the correspondence theory involves postulating the existence of at least two nonphysical entities—prop-

20. See John Feinberg, "Truth: Relationship of Theories of Truth to Hermeneutics," in *Hermeneutics, Inerrancy, and the Bible*, ed. Earl Radmacher and Robert Preus (Grand Rapids: Zondervan, 1984), pp. 3–50.

21. For an analysis of what makes our propositions true when they are true, see Kevin Mulligan, Peter Simons, and Barry Smith, "Truth Makers," *Philosophy and Phenomenological Research* 44 (1984): 287–321.

ositions and the relation of correspondence itself—a solution I favor, then one of the major ways of understanding science is undercut. That way of understanding science is contemporary physicalism, according to which all that exists and is causally efficacious in the world is matter and its modifications, and it leads to scientism.

It should be pointed out as well that there are alternative theories of truth that, if true, undercut a scientific realist understanding of science. Two of these are the coherence theory and the pragmatic theory. Put very simply, the coherence theory denies that individual propositions are true in virtue of their correspondence with a mind-independent world. Rather, truth is a property of an entire system of propositions when these propositions fit into a coherent web of beliefs. The pragmatist theory states that a proposition is true if and only if it "works." Much more could be said about each of these views, but enough has been said for our purposes.[22] The main point is that scientific realist understandings of science usually assume some form of correspondence theory of truth, and this theory involves a number of internal philosophical issues and faces the challenge of alternative understandings of truth. These issues are philosophical in nature. I am not suggesting that science cannot be justified until all problems about the nature of truth are solved. What I am suggesting is that if we cannot even begin to offer an analysis of the coresppondence theory that is sufficient both to clarify the theory itself (at least to some degree) and to ward off competitors, then the use of correspondence as a theory of truth by science is unjustified. Furthermore, the whole issue of competing views of truth is philosophical and presuppositional to science.

The Reliability of the Senses and the Mind

Phenomenology is a school of philosophy whose roots are in the ideas of the German philosopher Edmund Husserl. Phenomenologists emphasize the primacy of the knowing subject for all forms of relating to the world. We can only sense and reflect about the world from the vantage point of our own selves as first-person knowing and experiencing subjects. Thus, analysis and vindication of the different modes of experience—thinking, sensing, wishing, fearing, and so on—are prior to the various specific sciences that limit their reflection to specific

22. For more on coherentist and pragmatist theories of truth, see the sources in note 19.

areas of study like chemical phenomena, living phenomena, and the like.[23]

One does not need to agree with everything the phenomenologists say to acknowledge the primacy of assuming the reliability of our experiential faculties. Science assumes that our senses are reliable guides in knowing the external world and that our intellects are reliable guides in conceptualizing phenomena about the external world. Theories exist, including scientific theories. They are the products of thinking and sensing selves, and their reliability depends, in part, on the reliability of the senses and the intellect.

It is true that science itself analyzes different aspects of sensing and knowing. Neurologists study the brain and its activities, scientists study the physiology and psychology of sensory processes, and evolutionary theory seeks to give an account of how our faculties came to be what they are. But there is a difference between scientific and philosophical ways of approaching the intellect and the senses. For one thing, it is not at all clear that the mind and its processes are the same as the brain (and central nervous system) and its processes. Neither is it clear that experiencing the world in, say, sensory ways is merely a matter of passively receiving causal stimulation on the retinae, which in turn cause such-and-such physical processes to occur in the brain. It seems more likely that the eyes do not see anything, but that persons see by using the eyes. Regardless of whether one agrees about these matters, it is still true that philosophy asks and seeks to answer normative questions about the senses and intellect, and science asks and seeks to answer descriptive questions about the senses and intellect. Philosophy *seeks to justify* the deliverances of the mind and senses and theorizes about different normative theories regarding them, often in light of skeptical charges. Science *presupposes* the reliability of the mind and the senses and goes on to describe different aspects about the physical basis of their operation and their development and origination. After all, scientists justify their scientific models about the senses and the intellect by appealing to reasons and observations as normative courts of appeal for their theories. It would be question begging to use a scientific theory about the senses to justify the senses themselves. That justification is a philosophical matter, and the reliability of the mind and senses is a philosophical presupposition of science.

23. For an introductory treatment of phenomenological views of science see Richard M. Zaner, *The Way of Phenomenology: Criticism as a Philosophical Discipline* (Indianapolis: Bobbs-Merrill, 1970), pp. 41–78, especially pp. 51–62. For a more detailed discussion, see Aron Gurwitsch, *Phenomenology and the Theory of Science*, ed. Lester Embree (Evanston, Ill.: Northwestern University Press, 1974).

The Adequacy of Language to Describe the World

Scientific laws and theories are expressed in language. Science, therefore, is subject to philosophical disputes about the adequacy, use, and function of language. We will discuss these issues more in the next chapter. For now, I merely wish to point out that some thinkers have argued that language is inadequate to refer to and capture the way the language-independent world is in itself. On this view, language distorts reality. It reflects our own conceptual schemes about reality, not reality itself. Different language communities live in different worlds. Different scientific theories describe different worlds because the "world" each describes is not the world as it is in itself but the "world-in-the-theory." People in Alabama have one word for snow; Eskimos have twelve different words. Neither group is right; the two live in different worlds.

I do not wish to explore this dialectic further here. Suffice it to say that science presupposes that language is an adequate medium for referring to and stating truths about the world. Further, science assumes that different theories can be compared with each other against that world. These assumptions may appear common sensical, and I believe they are, for the most part, correct. Nevertheless, they are philosophical assumptions about the nature and function of language that are preconditions for science.

Mathematics and the Existence of Numbers

Science often uses mathematical language to describe the world. Now, such a use of mathematical language seems to presuppose that mathematical language is true, and if mathematical language is true, there must be mathematical entities called numbers that mathematical language refers to and truly describes. Put somewhat differently, some have argued that the only known account of mathematical truth not subject to serious difficulties is the same account of truth that works best outside of mathematics, namely, some form of the correspondence theory of truth. But if this is granted, then a sentence like "there are prime numbers greater than seventeen" is true only if there is some mathematical entity that exists and has these properties: being prime, being a number, and being greater than seventeen.

In order to understand this argument, let us look at what kind of ontological commitments seem to be part of fairly commonplace affirmations. Consider these sentences:

P: The apple is red.
Q: The apple is to the right of the desk.

Sentences P and Q can be reformulated as follows:

P': ($\exists$x) (Red x)
Q': ($\exists$x) (xRb) where R=to the right of, b=desk.

The paraphrase of P reads "There exists some entity, x, such that x is red." The paraphrase of Q reads "There exists some entity, x, such that x is to the right of b." Now, if P is true, then we are committed to 1) the existence of the apple referred to in the sentence and, some would argue, 2) the existence of the property redness had by the apple, and 3) some relation called exemplification (the apple exemplifies redness). If Q is true, then we are committed once again to the existence of an apple, a desk, and (perhaps) the spatial relation to-the-right-of. One thing seems clear—acceptance of the truth of P and Q carries with it acceptance of the existence of certain things, namely, an apple, a desk, and perhaps the universal called redness and the relations to-the-right-of and exemplification.

Now consider these sentences:

R: Two is an even number.
S: Nine is the smallest odd number greater than eight.

R and S appear to be true. But if so they can be paraphrased in such a way that the ontological commitments underlying acceptance of them become evident.

R': ($\exists$x) (Even x & x=2)
S': ($\exists$x) (xRb) where R=smallest odd number greater than, b=eight.

R involves at least the acceptance of some number identical to two that has the property of being even. S involves at least the acceptance of the number nine and the number eight. Now, it should be clear that, for example, the number two is not the same thing as the various numerals or symbols used to refer to the number two: "two," "zwei," "II," "2," and so on. These are physical scratchings on a sheet of paper and they all differ from one another, but the number they all refer to is a real abstract entity—a number—that exists and grounds the truth of mathematical claims like R.[24]

This argument for the existence of mathematical entities can be put in a slightly different way that makes essentially the same point. Biologists study living things and make true claims about them, so there must really be living things. Chemists study chemical elements and make true claims about them, so there must really be chemical elements. Mathematicians study numbers and make true claims about them, so there must be abstract, mathematical entities.

In sum, since science often uses mathematical language in a way that assumes the truth of mathematics, and if the truth assumed in mathematics is to be understood in the same way that truth is involved in fields outside mathematics, then the truth of mathematics carries with it an ontological commitment to abstract, mathematical entities that ground the truth of mathematical claims. Thus, some argue, science presupposes the truth of mathematics and, therefore, the existence of abstract, mathematical entities.[25]

However, it is not just the existence of mathematical entities that science assumes. Science also presupposes that mathematical language is adequate for describing the world. Often, mathematical theories are developed by mathematicians, theoretical physicists, or cosmologists in the privacy of a study without regard to the world. Someone simply plays with theorems, symbols, and the like, and generates certain results. However, it sometimes turns out later that the mathematical theory, developed with no regard to the world, accurately pictures some range of phenomena. As physicist Eugene Wigner put it: "The miracle of the appropriateness of the language of mathematics for the formulation of the laws of physics is a wonderful gift which we neither understand nor deserve. We should be grateful for it and hope that it will remain valid in future research. . . ."[26] In sum,

24. For an overview of different ways of understanding what numbers are, see Reinhardt Grossmann, *The Categorial Structure of the World* (Bloomington, Ind.: Indiana University Press, 1983), pp. 293–323.

25. See Bruce Aune, *Metaphysics: The Elements* (Minneapolis: University of Minnesota Press, 1985), pp. 19–22; Keith Campbell, *Metaphysics: An Introduction* (Encino, Calif.: Dickenson, 1976), pp. 200–205.

26. Eugene Wigner, "The Unreasonable Effectiveness of Mathematics in the Natural Sciences," reprinted in *The World of Physics, Vol. 3: The Evolutionaary Cosmos and the Limits of Science*, ed. Jefferson Hane Weaver (New York: Simon and Schuster, 1987), p. 96. Pages 13–96 of this volume are devoted to the use of mathematics in science. See also Mark Steiner, "The Application of Mathematics to Natural Science," *The Journal of Philosophy* 86 (September 1989): 449–80; Felix E. Browder, "Mathematics and the Sciences," in *Minnesota Studies in the Philosophy of Science Vol. 11: History and Philosophy of Modern Mathematics*, ed. William Aspray and Philip Kitcher

both the existence of abstract, mathematical entities and the applicability of the language of mathematics to physical phenomena are philosophical assumptions that underlie much of scientific practice.

The Concepts of Formal Ontology

When human beings think or speak about the world they classify things in various ways—this is a shoe, that is a red ball, and so forth. Of course, scientists classify things in just the same way—this is an electron, that is an organism of such-and-such species, that is an equilibrium process, and the like. Now, our classificatory concepts come in various degrees of generality. For example, the following is a list of classificatory concepts given in ascending order of generality for classifying my wife, Hope: "the individual 'Hope,' a human, a mammal, a vertebrate, an animal, a living substance, a substance-thing." Here is a similar list for classifying the color of my last book, *Scaling the Secular City*: "light blue, blue, a color, a visible property, a property-thing."

Some of our classificatory concepts are the broadest ones we have and reflect the way the world is and how we think about the world. These concepts are implicit in all of our classifications, and they are called formal concepts.[27] Here are some examples of formal concepts: a thing, an object, a concept, an event, a fact, existence, identity, and, perhaps, causation. In science, for example, formal concepts tend to be structural or organizational, that is, they help us organize and classify phenomena broadly and in such a way that we have some conception about how to approach those phenomena.[28]

Recently, philosopher George Bealer offered a powerful argument to the effect that certain formal concepts are essential to science and are involved in necessary truths (those true in every possible world, i.e., truths that do not merely happen to be true but must be true and cannot be false) that can be known (or perhaps must be known) a priori and constitute presuppositional, philosophical limits to science.[29]

Bealer argues that scientists frequently make what are called nat-

---

(Minneapolis: University of Minnesota Press, 1988), pp. 278–92. For an application of the "miracle" of the applicability of mathematical language to a design argument for God's existence, see James E. Horigan, *Chance or Design?* (New York: Philosophical Library, 1979), pp. 117–21.

27. For a brief discussion of formal concepts, see D. W. Hamlyn, *Metaphysics* (Cambridge: Cambridge University Press, 1984), pp. 54–59.

28. Cf. Harre, *Philosophies of Science*, pp. 27–29.

29. George Bealer, "The Philosophical Limits of Scientific Essentialism," in *Philosophical Perspectives, Vol. 1: Metaphysics, 1987*, ed. James E. Tomberlin (Atascadero, Calif.: Ridgeview, 1987), pp. 289–365.

ural-kind identities: water is identical to H_2O, temperature is identical to mean kinetic energy, gold is identical to the element with atomic number 79, and so on. Such identities presuppose and use two general kinds of formal concepts: *category* concepts and *content* concepts. Examples of category concepts are concepts of stuff: compositional stuff, functional stuff, substance, quality, quantity, action, artificial, natural, cause, reason, person. Examples of content concepts are basic phenomenal (mental) concepts (pain, itchiness, tingling sensation) and basic mental relations (knowing, perceiving, deciding, loving). Let us limit our discussion to category concepts and see how they enter into and are presupposed by the typical kind-identities that scientists make.[30]

Suppose a chemist investigates water here on Earth, that is, the stuff we cook with, use for ice cubes, fish in, ski on, and so forth, and discovers that all samples of it are composed of H_2O. So she concludes that, here on Earth, all and only samples of water are H_2O. Now suppose she visits another planet called Twin Earth and discovers samples of stuff macroscopically indistinguishable from water on earth (it looks, feels, smells the same and so forth), she finds that the inhabitants on Twin Earth cook in the material and use it for ice cubes, fishing, skiing, and so on. But suppose further that these samples are chemically composed of XYZ and not H_2O. Since water is identical to stuff composed of H_2O, then our chemist would conclude that the samples on Twin Earth were not samples of water, appearances to the contrary.

Now, Bealer continues, suppose we went to England to sample its cuisine, and we knew that at a certain time, the only food in England was mutton stew and all mutton stew in England was food. Mutton stew is composed of boiled mutton, potatoes, and turnips: MPT. Now imagine people from England visiting some location called Twin England where they discover fettucini, veal scallopini, and bread—things not composed of MPT—being cooked, eaten, served at restaurants, and so forth. Should these travelers call home to England and say "Lots of tasty, nutritious things to eat here, but sorry, no food"? After all, there is no MPT in Twin England. No, they should not. But, asks Bealer, how is this case different from Twin Earth?

The answer is that, unlike water, which is a compositional stuff, food is a functional stuff, and the identity conditions for compositional stuff are different from those appropriate for functional stuff. The iden-

30. In ibid., Bealer primarily discusses category concepts. He treats the presuppositional nature of content concepts in "The Logical Status of Mind" in *Midwest Studies in Philosophy 10: Studies in the Philosophy of Mind*, ed. Peter A. French, Theodore E. Uehling, Jr., and Howard K. Wettstein (Minneapolis: University of Minnesota Press, 1986), pp. 231–74.

tity conditions for compositional stuff involve the sample in question possessing the same composition as the material it is being compared with, while the identity conditions for functional stuff involve the sample in question serving the same function as the stuff it is being compared with. On Twin Earth, XYZ is functioning as water does on Earth, but it is not composed of H_2O; thus, being a compositional stuff, XYZ doesn't count as water. In Twin England, fettucini is not composed of MPT, but it does function as mutton stew does in England; thus, being a functional stuff, fettucini counts as food.

When scientists classify their objects of study, they assume various positions about formal concepts—category concepts in the preceding cases—and these assumptions determine how they will make their identities. Different category concepts entail different kinds of identities. But how do we know which formal concepts, or category concepts in this case, are appropriate? There is no scientific way to answer this question, since any scientific classification will itself presuppose some solution to the question about formal concepts. The appropriate way to take formal concepts is a matter of philosophical debate, and science presupposes solutions to this debate.

As an application of this point, Bealer asks about the problem of defining life.[31] All living things on Earth are composed mostly of H_2O plus certain macromolecules composed of various amino acids. However, this would not lead us to conclude that inhabitants on Twin Earth were not living if they had a radically different composition. The decision to count them as alive involves a philosophical decision about taking life as a functional concept and not a compositional one. I am not saying that this decision is arbitrary. Indeed, I don't think it is. But I am arguing, with Bealer, that this example illustrates that formal concepts are 1) essential to the kind-identities that science makes, and 2) philosophical presuppositions to science. Each of the formal concepts listed at the beginning of this section has been debated by philosophers, and various views are currently given about each. One's acceptance of certain understandings of formal concepts will affect how he understands the way these concepts work in all his intellectual life, including science.

The Existence of Values

Science is practiced by individuals who are members of different scientific communities—the communities of particle physicists, or-

---

31. For another application that involves the identity of various electrical phenomena, see Shapere, "Scientific Theories and Their Domains," pp. 518–25.

ganic chemists, entomologists, and so forth. The practice of science presupposes various individual and social values. Science is descriptive; it focuses on what is. Values involve what is prescriptive; they have to do with what ought to be. Even when scientists (e.g., sociologists) study values, they do so in a descriptive way. Sociologists cannot prescribe what is worthwhile in a culture, at least not as sociologists. They merely describe what is in fact viewed as worthwhile, whether marital fidelity or child sacrifice. Science describes and seeks to explain and interpret the various entities, relationships, and processes it investigates. It does not prescribe values.

Some values lie outside of science and will be discussed later. However, other values are preconditions for science itself. These values can be grouped into four broad classes: moral, aesthetic, epistemic, and methodological.

Individual scientists and communities of scientists assume various moral values that are part and parcel of experimentation, peer review, reporting of data, and so on. Some of the moral values assumed by science are accuracy of recording data, honesty in reporting one's findings to others, organized skepticism ("I'll have to wait to see the data"), universalism (i.e., scientific activities like journal acceptance should be governed by an impartial, impersonal code), communism (scientific data and information should be shared equally with all), unbiased objectivity and openness to refutation, and, perhaps, institutionalized rewards and punishments for those in the community who practice science correctly or incorrectly.[32] Without these and other moral values, science could not be done. But the existence, nature, and justification of moral values are matters of study for philosophy and theology, not science.

Scientists often use various aesthetic values as well. They often are guided by aesthetics and intuition, a sense of rightness regarding how

32. For more on the relationship between science and moral values, see H. Tristram Engelhardt, Jr., and Daniel Callahan, *Morals, Science, and Society* (Hastings-on-Hudson, N.Y.: The Hastings Center, 1978), pp. 1–155. William Broad and Nicholas Wade offer a study of various examples of fraud and deceit in science in *Betrayers of the Truth: Fraud and Deceit in the Halls of Science* (New York: Simon and Schuster, 1982). They conclude: "Science may be in one sense a community, but in another, equally important, it is a celebrity system. The social organization of science is designed to foster the production of an elite in which prestige comes not just on the merits of work but also because of position in the scientific heirarchy. Members of the scientific elite control the reward system of science and, through the peer review system, have a major voice in the allocation of scientific resources" (p. 214). One wonders if this conclusion helps to explain the force of the emotionalism and the use of ad hominem arguments against scientific creationists. Clearly, more is going on in the creation science debate than the objective rejection of an (allegedly) inadequate paradigm.

some phenomenon "has to be." Aesthetic considerations can be a mode of discrimination, a guideline for appropriateness in scientific expression. As Judith Wechsler has put it:

> Scientists talking about their own work and that of other scientists use the terms "beauty," "elegance," and "economy" with the euphoria of praise more characteristically applied to painting, music, and poetry. Or there is the exclamation of recognition—the "aha" that accompanies the discovery of a connection or an unexpected but utterly right realization in art and science. These are epithets of the sense of "fit"—of finding the most appropriate, evocative and correspondent expression for a reality heretofore unarticulated and unperceived, but strongly sensed and actively probed. The right formalism or model which "captures" this reality seems almost magical in its potency. Both art and science evoke the previously ineffable in making ideas clear, cogent, and manipulable.[33]

Aesthetic considerations sometimes are involved in the process of scientific investigation, conceptualization, and model making, as well as in the evaluation and acceptance of a scientific law or theory. Physicist Werner Heisenberg once said to Einstein:

> You may object that by speaking of simplicity and beauty I am introducing aesthetic criteria of truth, and I frankly admit that I am strongly attracted by the simplicity and beauty of the mathematical schemes which nature presents us. You must have felt this too: the almost frightening simplicity and wholeness of the relationship, which nature suddenly spreads out before us. . . .[34]

Along a similar vein, Paul Dirac wrote: "It is more important to have beauty in one's equations than to have them fit experiment. . . . It seems that if one is working from the point of view of getting beauty in one's equations, and if one has really a sound insight, one is on a sure line of progress."[35]

I am not suggesting that aesthetic considerations are the only ones relevant to science; far from it. Neither am I suggesting that they always are appealed to or that they always should be appealed to. I am suggesting that aesthetic considerations have played an important

33. Judith Wechsler, ed., *On Aesthetics in Science* (Cambridge, Mass.: MIT Press, 1978), p. 1.

34. Ibid.

35. Cited in Paul Davies, *God and the New Physics* (New York: Simon and Schuster, 1983), pp. 220–21.

role in the development of science, and when they have, science has presupposed aesthetic values.

Science also presupposes epistemic values. We have already delineated various epistemic values in chapter 2: simplicity, predictive success, empirical accuracy, internal clarity, and so on. These values are epistemic presuppositions of science, for they seek to spell out what epistemic virtues a "good" theory ought to embody. But the notion of an epistemic virtue and the distinction between the various epistemic virtues and their applicability to theories in general (and scientific theories in particular) both involve philosophical elucidation and debate.

Finally, science uses various methodological values. These can be classified roughly into two groups. First, there are methodological values involved within science, such as "prefer double-blind to single-blind experiments," "prefer results calibrated by instrument x over those calibrated by instrument y," "accept test results only after they have achieved such-and-such percent agreement," and so on. These methodological values involve standards of acceptance for scientific procedures, methods, instruments, and experimental design.[36]

A second class of methodological values focuses on science itself and involves standards for what science ought or ought not to be. For example, there was a time when Newtonian physics and atomism were so widely accepted that something was not counted as good science if it did not explain things in terms of the laws and principles of mechanics. Thus, biology was not regarded as good science.[37] Today, many scientists do not accept creation science because it refers to God as a theoretical, explanatory concept, and such a practice is supposed to be bad science. We will evaluate this claim later, but for now it is enough to point out that methodological standards for what is and is not good science are presuppositions of science itself—presuppositions that change from generation to generation and usually reflect, to some degree, the broader cultural context. Mechanistic atomists, Hegelians and other holists, Marxists, Christians and secularists, determinists and libertarians, dualists and physicalists, all may have different views about what is appropriately called good science. Of course, they may agree about what counts as good science, but when differences arise, they are partially a product of different methodological values about

36. Cf. Larry Laudan, *Science and Values: An Essay on the Aims of Science and Their Role in Scientific Debate* (Berkeley: University of California Press, 1984), pp. 23–41.

37. For an example of this type of thinking, see John R. Platt, "Strong Inference," *Science* 146 (October 16, 1964): 347–53.

science itself, and resolving or adjudicating those differences is a philosophical, not a scientific, task.

Singularities, Ultimate Boundary Conditions, and Brute Givens

Some features of the universe are brute givens from a scientific
point of view. That is, they are just there, and science uses them to
explain other things while leaving the givens themselves scientifically
unexplained. Science requires the presence of such givens, which thus
constitute a necessary precondition for science.

In order to illustrate a brute given, suppose we were trying to predict or account for the velocity v of some projectile at some time t
after its initial take-off. In order to determine the velocity v of the
projectile at t we need to know its initial conditions and the laws
governing its motion. Assuming an ideal condition with no friction,
low velocity, and so on, we can use one of Newton's laws of motion
for our calculation: $v = v_0 + at$, where v is the velocity we wish to cal-
culate at time t, v_0 is the initial velocity imparted to the projectile,
and a is the rate of acceleration of the projectile. The velocity of the
projectile during its flight is a function of this particular law and the
initial velocity. But the existence of this law and the particular value
of v_0 are brute givens as far as calculating v is concerned. The value of
v is a function of v_0 and this particular law, but they are simply taken
as givens.

Now it may be that v_0 and Newton's law of motion can be explained
in terms of prior conditions or subsumed under broader laws. For
example, v_0 may have the particular value it does because of the mass
and velocity of some other projectile that collided with the projectile
we are investigating at time $t = 0$. But sooner or later, at least as far
as science is concerned, we will arrive at various kinds of givens that
simply exist for science. From a scientific point of view, they are just
there, and they provide the starting point for scientific investigation.
Thus, science cannot achieve complete explanation of the universe
because of the presence of data inputs that are left scientifically
unexplained.

For example, a cosmic singularity is a given, an edge or boundary,
an initial condition where the laws of physics are undefined. The existence of the initial state of the Big Bang is the classic example of
such a singularity. It is a first cause, scientifically speaking. Philosophers may ask what caused the Big Bang, but if this state of affairs

was truly the first event of the space-time universe, then it is a given for science.[38]

Another example of a given for science is a cosmic constant or fundamental physical magnitude.[39] These are the particular values for various aspects of the universe that most likely could have been otherwise, but their specific values are givens for science, like the rate of expansion of the Big Bang, the speed of light, the mass of a proton, the various values of the forces of nature like the gravitational constant G.

Other examples of brute givens for science are the existence and nature of ultimate particles (it is a philosophical question as to whether there are such things as ultimate particles,[40] but if there are, as I believe, then their existence and properties are brute givens for science), ultimate laws of nature,[41] and the free acts of moral agents.[42] Regarding the last, if there are truly free, libertarian agents (and this is itself a philosophical question), then there is no set of prior conditions that is sufficient to cause an agent to act. An agent who chooses to do A and not B does so as a free agent who exercises his own causal powers as an agent in doing A. Such an act has the actor as a first cause, that is, his acting is necessary for the act to take place and such a free act is a given for science. It cannot be completely subsumed under prior states and laws of human action that are sufficient to explain the act.

In sum, givens for science serve as data inputs or ultimate laws that science uses to explain other things but are themselves left unex-

---

38. Some philosophers and scientists argue that there is an explanation for the initial conditions in the Big Bang, in which case those conditions are not givens. The explanation they have in mind appeals to future states of the universe, e.g., the initial conditions existed in order to bring conscious life about. I do believe that appeals to future, final causes are legitimate in science. But not all agree with me, and the role of final causes is itself a matter of philosophical debate that illustrates the presuppositional nature of philosophy for science. Furthermore, such an appeal to final causes may have the result of removing the Big Bang from the list of brute givens for science, but it does so by postulating another brute given, namely, the specific final cause or future goal the cosmos is moving toward. The existence of this goal and not another one would be a given for science.

39. For a list of these, see P. C. W. Davies, *The Accidental Universe* (Cambridge: Cambridge University Press, 1982).

40. See Richard Sorabji, *Time, Creation, and The Continuum* (Ithaca, N.Y.: Cornell University Press, 1983), pp. 321–421.

41. Richard Swinburne has used the existence of such ultimate laws for science as material for a design argument for God's existence. See his "The Argument from Design," *Philosophy* 43 (July 1968): 199–212.

42. William Rowe, "Two Concepts of Freedom," *Proceedings and Addresses of the American Philosophical Association*, Supplement to vol. 61, no. 1 (September 1987): 43–64.

plained by science. These givens are necessary initial conditions for science to function. Thus they constitute a precondition for the scientific enterprise, and our understanding of them is philosophically, not scientifically, arrived at.[43]

This completes our survey of presuppositions that underlie science. Let us now turn to a brief look at cognitive issues outside science.

Cognitive Issues Outside Science

Certain cognitive issues do not constitute preconditions for science but lie outside the domain of scientific investigation. These are cognitive issues in that they are not mere matters of subjective opinion but are amenable to rational argumentation and assessment, and claims about them can, at least in principle, be true or false. There are two broad classifications of cognitive issues outside science: extraneous cognitive issues and external conceptual problems.

Extraneous Cognitive Issues

Assessments and assertions about these issues do not clearly interface with science; thus, they and science are irrelevant to each other. Issues of this type are almost too numerous to mention, but the following list gives a feel for them: In theology, certain exegetical issues related to particular biblical texts are irrelevant to science. For instance, does John 3:16 refute a limited-atonement view of Christ's death or is it compatible with such a view? Certain more general interpretive matters also are irrelevant to science. For instance, what is the main purpose of the Book of Philippians? Some specific theological problems (e.g., whether Christ is declared to be God in the New Testament) and some debates between entire theological systems (e.g., whether dispensationalism or covenant theology is more adequate) are also irrelevant to science.

Some issues in history (e.g., establishing when the Pharaohs ruled, determining what caused Babylon to fall, and evaluating Lincoln's presidency), philosophy (e.g., debates between utilitarians as to what utility means and what version of utilitarianism is best), and many other disciplines involve rational argument, presentation of evidence, assessment of data, and the possibility of achieving truth (at least someone has the truth, whether or not we can agree on who that is;

43. Cf. Rescher, *Limits of Science*, pp. 174–218.

e.g., John 3:16 either is or is not compatible with limited atonement) but are irrelevant to science.

Two words of caution are in order. First, because of the kind of naive compartmentalization I criticized in chapters 1 and 2, some people tend to classify cognitive issues as extraneous too quickly. For example, Maxwell was metaphysically encouraged to view light not as a series of atomistically separate particles standing in external relations to other particles, but as fields whose various components (taken as a whole) are internally and intimately related to each other (change one aspect of the field and other aspects change as well). His metaphysical picture was grounded in two assumptions: that God is a Trinity wherein each member is internally related to the others in the single whole, and that God created his world in such a way that various aspects of it, light included, reflect his triune nature. At first sight, the doctrine of the Trinity seems utterly irrelevant to science, and Maxwell might have turned out to be wrong (he still might). But that is not the point. The point is that his metaphysics, grounded in trinitarianism, interacted with his science in helpful and appropriate ways.

Again, Aristotle and Plato disagreed over whether universals—entities like redness, triangularity, and humanness, which can be in more than one thing at the same time—can exist when all the particular instances of them cease to exist. Can redness continue to exist when all the red things (e.g., red balls) are destroyed? What is the relationship between a universal and its instances? These debates appear to be extraneous cognitive issues. But it would be false to conclude that all aspects of the problem of universals are irrelevant for science. Some have argued that if there are no universals, then there are no natures that science can discover and that ground the unity of classes of scientific entities. If there is no universal nature to an electron, then for all we know some electrons may not have negative charge.

Yet again, Copernicus was uncomfortable with realist interpretations of Ptolemy's model of planetary motion because it departed from the Platonic ideal of uniform circular motion about a center. According to him, motion that is circular is more perfect, and the planets seek to move in circular orbits. The point is not that Copernicus was right about specifically identifying circular orbits with perfection. But he was right in taking a rationally justified metaphysical framework (there were good arguments for that framework in his day) derived from Platonic views of universals and using it in his science. After all, scientists still talk about the world seeking to realize ideal energy states (e.g., minimum energy or maximum charge distribution in the transition form of an organic reaction).

The actual history of science shows a complex interaction among philosophy, theology, and some other fields in forming, testing, and using scientific theories. Such interactions often have been appropriate and fruitful. So what may appear to the modern reader, unacquainted with this history, as extraneous to science may on further inspection be something that can interact with it.

A second word of caution is that because scientism is so embedded in our culture, it often is assumed that if extraneous cognitive issues do not admit of precise solutions or solutions that all parties to the dispute agree on, or do not permit resolution by a method resembling scientific methodology, then such issues are not really cognitive. Apart from the self-refuting nature of such a claim (after all, the claim itself is not scientific), different areas of science admit of greater and lesser degrees of precision, depending on the applicability of mathematics to the problems in question, the level of the scientific problem (a low-level generalization versus a broad research program like atomism), the maturity and longevity of the branch of science itself, the acceptance of similar epistemic values by the disputants, the presence, force, and relevance of external conceptual problems, and so forth. Furthermore, some of these extraneous cognitive issues do use scientific methodologies in their resolution (e.g., exegetes form and test hypotheses and seek to explain various data in the texts in dispute). Finally, we have just seen that science is a first-order discipline that requires several philosophical presuppositions for its own justification. Thus, scientism is a misguided philosophy.

External Conceptual Problems

We have already discussed these in chapters 1 and 2, so we need only mention them here. An external conceptual problem arises for a scientific theory T when T conflicts with some doctrine of another theory T', and T' and its doctrines are rationally well founded, regardless of what discipline T' is associated with. For example, major philosophical arguments against an infinite past can be used to support the contention that there was a first event.[44] Any scientific cosmology that postulates an infinite past (e.g., an oscillating universe) will conflict with these philosophical arguments and their conclusion, a fact that will tend to count against the infinite-past cosmology.

Different external conceptual problems will have different degrees of epistemic weight, and nothing helpful can be said about them in general. A given problem may simply count against a scientific theory,

---

44. See Moreland, *Scaling the Secular City: A Defense of Christianity* (Grand Rapids: Baker, 1987), pp. 15–42.

but its force may not be sufficient to justify suspension of judgment about that theory. Other external conceptual problems may be sufficiently telling to justify abandonment of a given scientific theory. Only a case-by-case investigation can, at least in principle, indicate what state of affairs is present regarding the assessment of some particular theory. What is important is that such problems have existed, do exist, and play an important role in science.

Limits within Science

Some issues are not preconditions of science and do not arise outside of science but arise from science itself or from some field like philosophy and limit the very practice of science.

For example, some limits arise on the basis of currently accepted scientific theories. These limits may only be temporary if in the future we abandon the theories that generate them. For example, "What goes on inside a black hole?" cannot be answered given our current beliefs about black holes. According to current theory, black holes by their very nature do not permit data extraction. A black hole is a gravitationally collapsed mass from which no light, mass, or any other kind of signal can escape. Again, Heisenberg's principle of uncertainty states that there are inherent limits as to how precisely we can determine simultaneously the position and momentum of a particle.

These limits come from scientific theories themselves. They are not limits in our understanding of a scientific theory or in our ability to extend or make it more precise. Rather, they are predicted limits based on ideas central to the scientific theories themselves. The limits may be only temporal, but if they are, it will be not because we refine but because we abandon the theories that generate them.

Nicholas Rescher has pointed out another internal limitation to science.[45] This limitation derives from the very nature of questions themselves, scientific questions included. In order for a question to make sense and not just be unintelligible gibberish, certain presuppositions of the question must be assumed to be meaningful. For example, "What is the melting point of lead?" makes sense only if we assume that there is such a thing as a melting point and we know what one is and how to recognize it, if lead is one of the things that possibly has a fixed, stable melting point, and so on. The presuppositions constitute the formative background that make the question possible. Without that background, the question cannot even arise.

45. Rescher, *Limits of Science*, pp. 18–27.

For example, Newton could not even have asked about the makeup of DNA or the quantum levels of the atom because the necessary conceptual apparatus for these questions to make sense was not present in his day. If quantum theory is true or approximately true, then quantum levels existed during Newton's time, but he could not have asked about them. Thus, the possibility of a question depends on the intelligibility of its presuppositions, and these in turn depend on and are relative to the current conceptual framework of science. Thus, some questions are limits for science not because we understand them but are ignorant of their answers but because we cannot even pose them. Again, these types of limits can be removed as science grows and expands, but new ones are always present relative to the current scientific framework.

A third limit within science arises from another observation about questions.[46] This observation has been called the Kantian principle of question propagation: The answering of our factual scientific questions always paves the way to further unanswered questions. As we explain various phenomena, we discover other phenomena that require explanation. Stated truths express facts, but there are always facts that have not been stated. The more we come to know, the more there is to know about the system of which we ourselves are a part.

Finally, I mentioned in the introduction to this chapter that one branch of science can limit another. Chemists take the particular ensemble of chemicals in some organism as a given and describe their relationships, processes, and so on. But biologists explain the very existence of this ensemble instead of another one by appealing to biological concepts outside the range of chemistry: the history of the development of the organism, the functional role of its macroorgans (e.g., the heart, the kidneys), and the relationship between the organism and other features of its ecosystem. Unless one field of science is reducible to another (e.g., biology to chemistry and physics), this kind of limit will exist within science.

Summary and Conclusion

We have seen that scientism, or scientific imperialism, is a false philosophical view of science. It is a self-refuting philosophical claim. By using science to define science, scientism fails to understand adequately the various limits to science—presuppositions that underlie

46. Ibid., pp. 27–34.

science, cognitive issues outside science, and limits within science. As Rescher puts it:

> Science does not have exclusive rights to "knowledge:" its province is far narrower than that of inquiring reason in general. Even among the "modes of knowledge," science represents only one among others. It is geared to the use of theory to triangulate from objectively observational experience to answer our questions about how things work in the world. But there are many other areas in which we have cognitive interest—areas wholly outside the province of science.[47]

47. Ibid., pp. 214–15.

Scientific Realism

> ... if one wants a slogan: realism is the truth and temperate rationalism the way.
> —W. H. Newton-Smith, *The Rationality of Science*

> There is, I think, no theory-independent way to reconstruct phrases like "really there;" the notion of a match between the ontology of a theory and its "real" counterpart in nature now seems to me illusive in principle. Besides, as a historian, I am impressed with the implausibility of the view.
> —Thomas Kuhn, *The Structure of Scientific Revolutions*

Ever since the ancient Greeks began to investigate nature, there has been a debate between realist and antirealist understandings of science. The term *realism* has a wide variety of meanings in philosophy. It can be used in opposition to nominalism (a realist accepts, while a nominalist rejects, the existence of universals), to perceptual dualism (a realist believes we see objects in the world directly, a perceptual dualist believes we see only sense images directly), or to metaphysical idealism (a realist accepts, while an idealist rejects, the full reality of the physical universe). Very roughly, in the context of debates about science, scientific realism holds that science progressively secures true, or approximately true, theories about the real, theory-independent world "out there" and does so in a rationally justifiable way. Antirealism, in all of its stripes, denies realist interpretations of the scientific enterprise and offers various alternative understandings.[1]

1. For the history of the philosophy of science, see Stanley Jaki, *The Road of Science and the Ways to God* (Chicago: University of Chicago Press, 1978) and John Losee, *A Historical Introduction to the Philosophy of Science*, 2d ed. (Oxford: Oxford University Press, 1980). Developments in the philosophy of science in more recent years are surveyed in Larry Laudan, *Science and Values: An Essay on the Aims of Science and Their Role in Scientific Debate* (Berkeley: University of California Press, 1984), pp. 1–

Figure 4.1
Options in the Realist/Antirealist Debate

The major issues in the realist/antirealist debate involve questions like these: Do the theories of science give a literally true model of the way the world is, or do they merely provide useful fictions, calculating devices, or convenient summaries of sensory experience that "work" (e.g., help us control nature, predict phenomena, and so on)? When is it reasonable to accept a realist or antirealist understanding of science? Which account is better able to explain the success of science as a problem solver?

There are several different options in the realist/antirealist debate. Figure 4.1 represents one way of categorizing them.

We will use the term *rational realism* for scientific realism in this and the next chapter. *Rational nonrealism*, also called *instrumentalism*, comes in several varieties, but all agree that science is an objectively rational discipline (we ought to accept good scientific theories) but does not aim at giving us truth or approximate truth; that, instead, science seeks to do a variety of other epistemic functions (e.g., synthesize sense data, predict and control phenomena). *Nonrational nonrealism* also comes in different forms, but all adherents of this school accept at least two theses: science does not seek truth or approximate truth; there is no objective notion of rationality available that either sets science off from other disciplines as objectively rational or sets some particular scientific theory off as more rational than another.

In chapter 5 we will investigate different alternatives to rational realism and clarify further the details of these alternatives. All three schools of thought have able representatives, and there is no small debate today about the realist/antirealist question.

In this chapter we will state rational realism in more detail and view some of the criticisms of it. We will not investigate the positive

22, and David Braybrooke and Alexander Rosenberg, "Comment: Getting the War News Straight: The Actual Situation in the Philosophy of Science," *The American Political Science Review* 66 (Sept. 1972): 818–26. For an analysis of the different ways of relating the philosophy of science and the history of science, see Losee, *Philosophy of Science and Historical Enquiry* (Oxford: Oxford University Press, 1987).

arguments in support of it, at least not directly, for three reasons. First, one of the major arguments for rational realism is that it accounts for the success and progress of science vis-á-vis other disciplines better than any other understanding of science. This claim will be investigated both in analyzing the nature of scientific progress and in looking at other accounts of the scientific enterprise in chapter 5. Second, other arguments for rational realism involve refuting arguments against it and presenting alternatives to them, so we will get at those issues by investigating some of those criticisms themselves. Third, rational realism is clearly the majority view among scientists and those who make claims about the relationship between science and Christianity, so the task of exposing its weaknesses is sufficiently pressing to warrant focusing on them.

My own eclectic view incorporates a chastened form of rational realism. I am eclectic because I think the criticisms of rational realism and the strength of some of the alternatives to it are sufficient to warrant adopting a realist or an antirealist attitude toward some particular areas of science on a case-by-case basis. For example, it may be best to adopt a realist attitude toward certain areas of geology but an antirealist attitude toward certain areas of theoretical physics. The reasons for my eclecticism should become clear in this and the next chapter. Further, when I do adopt a rational realist perspective about some particular area of science, I am not as certain about my realism as I used to be, so even when I wear a realist stripe, I do so with varying degrees of tentativeness. Again, the reasons for this tentativeness should become evident in this and the next chapter.

Why worry about the realist/antirealist debate? What relevance does it have for those who are attempting to integrate science and theology? In order to be a fully actualized and integrated human being and a mature Christian with no secular/sacred dichotomy, one needs a coherent, intellectually satisfying Christian world view. Such a world view involves, among other things, fitting science and theology together in a harmonious way. Here is where the realist debate is relevant.

A realist understanding of science can contribute to an approach to the theology/science dialogue by implying that if science says something is true and rational and Christian theology seems to conflict with it, then Christianity needs to be adjusted in some way. The result of this has been the idea that since science always defeats religion in battle—itself a suspect claim—then perhaps religion was never intended to be a factual, rational way of understanding the world of nature, but is merely a private guide for practical life or a guide to truth in "spiritual" matters.

If, however, there is reason to doubt that scientific theories are rational or approximately true, even if they are "good" theories, then it is difficult to see how science could pose a threat to the truth or rationality of Christianity.[2] Most debates between science and theology (e.g., the creation/evolution controversy) assume rational realism. After all, if one does not assume that evolutionary theory is approximately true or objectively rational, then why bother to argue against it or to integrate Christian theological claims about the history of life with evolutionary theory? If quantum physics denies the law of cause and effect if understood in realist terms, this presents (not necessarily insuperable) problems for arguments for God based on causation. But if quantum physics is understood in antirealist terms, then it may be irrelevant for cosmological arguments that appeal to the universality of the law of cause and effect.

We will look first at what rational realism is and then investigate some of the criticisms of it.

What Is Rational Realism?

Rational realism was a minority view among philosophers of science in the first half of this century, but it currently enjoys majority status among both scientists and philosophers of science. Prominent rational realists are Karl Popper, Richard Boyd, W. H. Newton-Smith, Rom Harre, Ernan McMullin, and Hartry Field. There are different varieties of rational realism, but its core tenets are these:[3]

RR1. Scientific theories (in mature, developed sciences) are true or approximately true.

RR2. The central observational and theoretical terms of a mature

2. At least not the kind of threat involved in a realist understanding. Some forms of antirealism have been extremely empiricistic, i.e., the meaning of our words consists in how we empirically verify them, so unobservable theoretical entities like electrons and unobservable beings like God are ruled out as subjects of meaningful discourse. This type of logical positivism is seldom embraced these days (for one thing, it is self-refuting), and in any case, the view popular among many today—that science is true and rational and theology is a matter of belief and meaning, not of facts and rationality—involves a realist understanding of science.

3. See Laudan, *Science and Values*, pp. 104–9; Jarrett Leplin, ed., *Scientific Realism* (Berkeley: University of California Press, 1985), pp. 1–7; Del Ratzsch, *Philosophy of Science*, ed. C. Stephen Evans (Downers Grove: Inter-Varsity, 1986), pp. 80–90; W. H. Newton-Smith, *The Rationality of Science* (Boston: Routledge and Kegan Paul, 1981), p. 43; Nicholas Rescher, *The Limits of Science* (Berkeley: University of California Press, 1984), pp. 153–59; Rom Harre, *Varieties of Realism: A Rationale for the Natural Sciences* (Oxford: Basil Blackwell, 1986), pp. 34–95.

scientific theory genuinely refer to things—entities, processes, and so forth—in the world. These terms make existence claims.

RR3. It is possible in principle to have good reasons for thinking which of a pair of rival theories is more likely to be true or closer to the truth. Rationality is an objective notion and conceptual relativism (which holds that what is rational for one person or group should not necessarily be so for another person or group since rationality is itself relative to a person, community, or theory) is false.

RR4. A scientific theory will embody certain epistemic virtues (simplicity, clarity, internal and external consistency, predictive success, empirical accuracy, scope of relevance, fruitfulness in guiding new research, etc.) to greater and lesser degrees proportionate to its truth or degree of approximate truth.

RR5. The aim of science is a literally true conception of the world. Scientific progress is a fact, and science tends to converge on truer and truer conceptions of the world, with later theories usually refining and preserving the best parts of earlier theories and coming closer to the truth than earlier theories in the same domain of discourse (i.e., where the earlier and later theories are talking about the same things and thus are comparable to each other).

The Correspondence Theory of Truth

RR1 expresses the conviction that science uses a correspondence theory of truth: a theory is true if and only if what it says about the world does in fact accurately describe the world. The theory-independent world is what makes our theories true. Furthermore, RR1 implies that the idea of approximate truth is coherent. Truth is a matter of degrees, a theory can grow in truth, and one theory can be more true than another.

The notion of a mature, developed science has two aspects. First, an immature area of science can become mature if it has time to develop, test, and "vindicate" a major paradigm, if it investigates a clear domain of entities, if it is not a recent branch of science, and so forth. This notion of maturity is somewhat slippery, especially if one tries to tighten it up, but it is fairly clear at an intuitive level. Second, some sciences are considered mature in the sense of being the ideal, archetypical sciences. In this sense, physics and, perhaps, chemistry are considered mature in that they are the ideal sciences.

The Relation of Language and Reality

RR2 is a semantic thesis about the nature of language or, more specifically, about whether and how terms refer to things in the world and get their meaning. Since scientific theories have meaning, refer to things (allegedly) in the world, and are stated in a language, a discussion of scientific theories will include issues relevant to discussion of language in general.

To illustrate this, consider the following two sentences:

1. Fido is brown.
2. The average family has 2.5 children.

In sentence 1, the term *Fido* is a referring term. It refers to the dog Fido, an extralinguistic entity in the world, and says of him that he has the property *brownness*. In sentence 2, the term *the average family* appears to function just as the name *Fido* does in sentence 1, at least if we compare the surface grammars of both sentences, since *Fido* and *the average family* both occupy the subject position in a subject-predicate sentence. However, on closer inspection we realize that *the average family* is not a referring term at all. No one would try to locate where the average family lives and count its children to see if, in fact, there are 2.5 of them. Rather, *the average family* is a shorthand term for a set of mathematical operations. It says, "Add the number of children and divide by the number of families and you get 2.5."

Now consider this sentence:

3. Protons have positive charge.

RR2 says that the theoretical term *proton* is a referring term. Thus, if *proton* is in an accepted scientific statement like statement 3, then this has ontological implications. Just as *Fido* implies the existence of the dog, Fido, if one accepts sentence 1, *proton* implies the existence of an atomic entity, a proton, if one accepts sentence 3. Extralinguistic entities, protons, exist according to a realist interpretation of sentences like 3.

RR2 also makes use of the notions of central observational and theoretical terms. We have already seen, in chapter 1, several criticisms of an absolute distinction between observational and theoretical terms, and these criticisms have been used against rational realism. Nevertheless, it does seem that a continuum exists between observational terms and theoretical terms, and the distinction is not without importance. Roughly speaking, then, we can understand the observation/theory (O/T) distinction as follows: An observational term (O-term) like *is red, points to five, sinks, is hard, has such-and-such volume,* and so on, describes or reports sensory observations. A theoretical term (T-term) like *zero rest mass, electron, kinetic energy, field, gene, temperature, atomic number,* and so on, refers to some theoretical entity postulated to play a role in some theory.[4]

What does it mean for an observational or theoretical term to be central to a theory? Suppose we were trying to identify some substance, X, that we think is copper.[5] Given our current theories about metals in general, and copper in particular, a number of theoretical properties or observational consequences are relevant for classifying X as copper: being reddish, having the melting point of 1083° C, and having the atomic number 29. Some of these are more relevant than others. With respect to two theoretical or observational properties, one is more central to classifying X than another if and only if the possession (or lack) of that property tends to count more in favor of (or against) classifying X as copper than does possession (or lack) of the other property.

With regard to copper, having the atomic number 29 is more important than melting at 1083° C, which in turn is more important than being reddish. The more central a property is, the more weight it has in settling a dispute about classification. The intuitive ideas here are that scientific theories have several theoretical concepts and observational consequences, that some of these are more important for the theory than others, and that those judged more important are the ones that are more central. Centrality is a matter of degrees, and what is or is not central can change with time. In the earlier days of chemistry, one of the chief ways of classifying a chemical was its method of preparation. But today this carries little weight; the same chemical compound can be prepared in a variety of ways.

In any case, RR2 claims that those observation or theory terms that are central to a theory genuinely refer to real entities in the world. Copper really does have 29 protons in its nucleus.

Science Is a Rational Discipline

RR3 asserts that science is an objectively rational discipline, not in the sense that scientists have no biases but in the sense that one can

---

4. Some terms can be both. Thus, "continental plate" is a theoretical term in continental drift theory, but it could also be observed, at least in principle.

5. See Peter Achinstein, *Concepts of Science: A Philosophical Analysis* (Baltimore: Johns Hopkins, 1968), pp. 1–46, especially pp. 19–21; Newton-Smith, *Rationality of Science,* pp. 19–43.

have objective reasons or warrant for accepting or rejecting a given scientific theory. Rationality is a normative, objective notion, not a relative one. Furthermore, when two rival theories are being judged (e.g., Newton's or Einstein's concept of space, time, or mass), there are, in principle, considerations that can be used to judge between them. The theories are commensurable; they can be compared vis-á-vis each other against some common ground—the data, theoretical simplicity, and so forth.

Epistemic Grounds for Truth Judgments

RR4 claims that if a scientific theory has certain epistemic virtues, then it is objectively rational to believe that it is true or approximately true *because* it has these virtues. If T_1 is simpler than T_2 or explains more data, gives more accurate predictions, and so on, then one should believe that T_1 is truer than T_2. Furthermore, if a theory is true and its rivals are false, or if a theory is truer than its rivals, then one should expect it to be better eventually at predicting data, containing clarified terms, or embodying other epistemic virtues.

Science Aims at Truth

Finally, RR5 states that the aim of science is to give us not just theories that work (e.g., that help us control nature and predict phenomena), but theories that are true. Science tries to tell us the way the world really is, and more recent scientific theories are more accurate pictures of the world than are their predecessors. Science progresses over time toward a true picture of the world. Scientific progress is to be understood as progress toward truth.

A good deal more could and perhaps should be said about rational realism, but this is a good overview of it.

Criticisms of Rational Realism

Rational realism is the majority view about science embraced by philosophers of science and scientists alike. Nevertheless, it has not always been so popular. In the first half of this century, for example, it was in the minority. Even today some scholars do not find it an adequate philosophical understanding of science. In fact, it is safe to say that in the last twenty-five years there has been a consistent, broadly based, and intellectually sophisticated attack on rational realism. Let us take a look at some of the criticisms of this interpretation of science.

Truth, Approximate Truth, and Verisimilitude

We have already seen that several objections have been raised against the correspondence theory of truth, and while I still think that theory is preferable to all options, these objections do carry weight, especially when theories of truth are assessed in more theoretical contexts outside of daily life, like science. The problems with the correspondence theory have persuaded a number of thinkers to abandon rational realism.[6]

First, as we saw in chapter 1, several philosophers have criticized the observation/theory distinction. Seeing something is not a passive matter of receiving stimuli on one's retinae. Rather, seeing involves seeing *as* or seeing *that;* it involves an interpretive element. One must have conceptual notions in place before he can recognize or classify what he sees. Without the concept of an electron or the color red already in place, one cannot count certain things as observations of these entities.

The dependence of observation on conceptualization, so the objection goes, is especially prominent in the more theoretical deliberations about the world exemplified in science. Our theoretical concepts strongly influence, perhaps determine, what we see and what we count as an observation. Regardless of whether this point is made along Kantian lines (our concepts determine and shape our perceptions) or along Berkeleyan lines as stated in chapter 3 (we never directly see the world but only see our sensory images or experiences), the idea of a mind- or theory-independent world for our theories to correspond to becomes problematic. Either the very idea of such a world is unintelligible, or even if such a world exists "out there," we cannot get to it, we cannot step outside our theories of the world to test them.

These difficulties with the correspondence theory have been thought to be especially problematic when we are dealing with unseen (and unseeable) theoretical entities to which our only access is through the very theories that postulate their existence. Thus even if one is convinced that these problems with the correspondence theory can be

6. The correspondence theory seems to me to be correct and is most in keeping with a biblical view of truth. See Norman Geisler, "The Concept of Truth in the Inerrancy Debate," *Bibliotheca Sacra* (October–December 1980): 327–39. The correspondence theory does, however, present a real problem for those rational realists who want to remain physicalists (the view that only matter and its modifications are real), because the best understanding of the correspondence theory uses nonphysical entities like propositions, concepts, and the relation of intentionality (or denotation). See Hartry Field, "Tarski's Theory of Truth," *The Journal of Philosophy* 69 (July 1972): 347–75, and Dallas Willard, *Logic and the Objectivity of Knowledge* (Athens, Ohio: Ohio University Press, 1984), pp. 166–86.

overcome in normal discourse about macro-objects (chairs, dogs, and the like), they may still be telling against the application of the correspondence theory to theoretical entities to which no direct, theory-independent access is possible. And even when we consider regular objects like dogs, when a scientist claims to observe such an object, she is not merely claiming to observe a brown, furry creature that barks and eats dog chow. She is claiming to see an object that satisfies an elaborate biological classificatory theory. Even here, then, direct observation may be difficult—or so say critics of the correspondence theory.

Second, there are difficulties in spelling out what the correspondence theory takes truth to be. What are the entities that correspond to the world? What in the world makes our theories about it true? For example, our theories about the world exhibit a logical structure. But does the world exhibit such a structure? And what is the relation of correspondence itself? Furthermore, the correspondence theory of truth seems to be grounded in a more basic relation, the relation of reference, which obtains between our terms or concepts and the world. "Fido is brown" is true only if at least the term *Fido* succeeds in referring to an extralinguistic entity, the dog in the living room. But there are several problems with unpacking an adequate account of reference, as we will see later in this chapter. Thus, both an analysis of the correspondence theory and an analysis of reference, which underlies the correspondence theory, are so problematic as to warrant abandonment of the correspondence theory.

These considerations have led some philosophers, like Hilary Putnam, to change the notion of truth from a metaphysical one to an epistemological one. Normally, truth has been thought of as a metaphysical idea: it involves clarifying what truth is and what the world itself is apart from our theorizing about it. But if such an idealized notion of the world, a notion that views the world as some ideal limit to our progressive theoretical approximations of it, is troublesome, perhaps truth should be defined in terms of our *knowledge* of the world.[7] Putnam defines truth as what we are warranted in asserting. Truth is identical to a good theory that fits data and embodies several of the epistemic virtues listed under RR4.[8] But this amounts to an abandon-

7. See Hilary Putnam, *Philosophical Papers, Volume 3: Reason, Truth, and History* (Cambridge: Cambridge University Press, 1981), pp. 49–74.

8. In epistemology, we wish to explain how it is that when our beliefs have features that make them justified, we claim that we have truth. The desire to hold truth and justification together is one reason why some philosophers embrace a coherence theory of truth and a coherence theory of justification (criteria for knowing that we have the truth). If truth simply is the coherence of our beliefs, then when our beliefs exhibit

ment of realism. According to realism, we may have the truth when our theories are warranted or when they exhibit good epistemic virtues, but truth is not the same thing as having those virtues.

In light of these and other criticisms, some philosophers have recommended that we abandon a correspondence theory of truth and, with it, rational realism. There is another set of problems involving truth and science that involves the notion of approximate truth. Almost no one would say that our current scientific theories are the whole truth and nothing but the truth. It is surely possible, indeed likely, that the science of 150 years from now will differ to some degree or other from the science of today. This observation has led rational realists to say that while we do not have truth in our past or present (and by extension, future) theories, we still have achieved approximate truth. Our current theories are truer than earlier ones, and future theories will be truer still.

The notion of approximate truth is called verisimilitude.[9] Verisimilitude refers to the extent to which a theory captures the whole truth. Theory A has greater verisimilitude than theory B if and only if A is more approximately true than B. Theories can increase in verisimilitude; that is, they can increase in the degree that they are approximately true. If the notion of verisimilitude can be adequately spelled out, then rational realists are claiming not that our current theories are the final truth of the matter but only that scientific progress is measured in growth in verisimilitude.[10]

However, a number of philosophers of science have pointed out that the notion of verisimilitude is inadequate.[11] Two problems are central to criticisms of verisimilitude. First, no one has stated adequately what approximate truth means. Something is either true or false. Truth does not come in degrees.[12] A false generalization that works better

---

such coherence, we simply have the truth. For more on this, see Jonathan Dancy, *Introduction to Contemporary Epistemology* (New York: Basil Blackwell, 1985), pp. 110–40.

9. Verisimilitude is not to be confused with high probability that beliefs are true. The latter means that our certainty is probable, but not the belief itself, which is completely true if it is true at all. Verisimilitude means that the belief itself is approximately true.

10. See Jarret Leplin, "Truth and Scientific Progress," in *Scientific Realism*, ed. Leplin, pp. 193–217.

11. See Laudan, *Science and Values*, pp. 117–24, and *Progress and Its Problems* (Berkeley: University of California Press, 1977), pp. 123–27; Harre, *Varieties of Realism*, pp. 35–51.

12. A coherence theory of truth can embrace degrees of truth, for such a theory sees truth as the degree of "coherent fit" among the propositions of a web of beliefs, and such a fit can grow or diminish. However, the coherence theory of truth is not embraced by rational realism.

than another false generalization is still false. For example, granting that we accept Einstein's laws of motion, space, time, and mass, Newton's laws of motion are literally false, even at very low velocities, even though at those velocities they allow us to do a number of things we are interested in: predict the location of a projectile, aim tank guns, and so forth. Again, suppose there were a circular sphere in another room unavailable to us. Jones theorizes that the object is elliptical and Smith conjectures that it is square. Literally speaking, both of them have false theories, even though the circle resembles an ellipse more than it does a square, and for some practical purpose besides truth (e.g., securing something to shoot baskets with) the elliptical hypothesis gives us more hope than does the square hypothesis. So something either is or is not true, and the idea of approximate truth is problematic.

Nevertheless, some philosophers have tried to make sense of verisimilitude. One of the first attempts to do this in the context of scientific progress in recent years was offered by Karl Popper.[13] According to Popper, if we compare two rival theories, T_1 and T_2, T_2 will be of greater verisimilitude than T_1 if these conditions obtain:

1. The truth content of T_2 is greater than that of T_1.
2. The falsity content of T_1 is greater than that of T_2.

The truth and falsity content of a theory are, respectively, the number of true or false sentences entailed by the theory. For example, two rival theories about the relationship between pressure, temperature, and volume of a gas may have a potentially infinite set of sentences correlating these three properties, and one theory could have more true sentences and fewer false ones than another. In this case, the former would be more approximately true.

Even if Popper's understanding of approximate truth is correct, and most do not think it is, it still fails to tell us adequately how we could ever be in the position of telling when one theory had greater verisimilitude than another. What are the criteria for such a judgment? It would be possible for T_2 to be more approximately true than T_1, but for T_1 to fare better than T_2 in light of scientific testing. It very easily could happen that all the tests we have done so far, or all the tests we could do for several years to come, would be tests surfacing the false propositions of T_2 and the true propositions of T_1. Remember, both theories have true and false propositions. Verisimilitude merely guarantees that T_2 entails more true propositions and fewer false ones than T_1. But we might have tested only the false propositions in T_2 and the

13. Popper, *Conjectures and Refutations: The Growth of Scientific Knowledge* (New York: Harper and Row, 1963), pp. 215–50, especially pp. 231–35.

true propositions in T_1. To make a judgment between T_1 and T_2 as to their Popperian verisimilitude, we would have to have a God's-eye view, seeing all the truths entailed by each theory and comparing their relative number. But we are not in such a position.

Some rational realists have responded to this challenge by giving criteria for how we can tell which of a pair of rival theories has greater verisimilitude. Popper himself offered six criteria for determining when T_2 has greater verisimilitude than T_1: T_2 makes more precise assertions, it explains more facts, it has passed tests that T_1 has failed, it suggests and passes new experimental tests, it unifies or connects more hitherto unrelated phenomena than does T_1.[14]

Unfortunately, Popper merely lists the presence of epistemic virtues in a theory as an indication of its verisimilitude, but this is precisely what the antirealist denies. Thus, such an appeal is question-begging when used against the antirealist, for on his view, a theory's embodying several epistemic virtues (e.g., empirical precision, passing tests, suggestion of new experiments, and so on) only proves that it works, not that it is true. As anitrealist Nicholas Rescher puts it:

> But why should we take the stance that the movement of scientific knowledge is substantively *directional*—that science may indeed not have reached the ultimate truth but is drawing nearer to it? The reply is that this idea of "drawing closer to a definitive picture of reality" rests on a fallacy. It proposes to move from the fact that we have better warrant for accepting a revised picture of nature to the conclusion that this picture is closer to the truth (more faithful)—that is, it moves from a picture-of-nature's being a *better warranted picture* to its being a *better picture*. And this is a slide we cannot make, for the later theories of science are superior to the earlier not because of their more faithful delineation of nature but because they afford us enhanced means of prediction and control.[15]

In sum, for these and other reasons, several contemporary philosophers and scientists have abandoned the notion of truth and approximate truth and, with them, rational realism. Scientific theories, they say, are neither true nor approximately true descriptions of the world.

Theoretical Terms Can Be Eliminated

Realists interpret the role of (at least most) theoretical terms as referring to theoretical entities.[16] However, it is often possible to elim-

14. Ibid., p. 232. See also Newton-Smith, *Rationality of Science*, pp. 183–207.

15. Rescher, *Limits of Science*, p. 73. See also Laudan, *Science and Values*, pp. 117–24.

16. See Karel Lambert and Gordon G. Brittan, Jr., *An Introduction to the Philosophy of Science* (Atascadero, Calif.: Ridgeview, 1970), pp. 146–58.

inate the theoretical terms altogether without loss of empirical content, that is, without loss of the ability to predict observations accurately. If this is the case, some argue, then the role of theoretical terms is to be mere instruments, convenient summaries of sensory data that allow one to predict other sensory data and extend the range of application of our empirical generalizations.

An example may clarify this. Some empirical generalizations are limited in their range of application and have exceptions, like the following:

> 1. Wood floats on water, iron sinks in water.

This generalization is limited to wood, water, and iron, and it has exceptions: ebony chips sink and iron ships float. But we can fix the situation by introducing a theoretical concept, specific gravity, defined as the quotient of a solid body's weight divided by its volume. Now our generalization can be modified to remove exceptions and extend its application beyond wood, iron, and water:

> 2. A solid body floats on a liquid if its specific gravity is less than that of the liquid.

If the theoretical term *specific gravity* is interpreted realistically, then it refers to some property of things in the world. However, *specific gravity* can be eliminated without loss of empirical accuracy or predictive success as follows:

> 3. A solid body floats on a liquid if the quotient of its weight and volume is less than the corresponding quotient for the liquid.

Sentence 3 reveals that "specific gravity" is a mere shorthand term for a set of calculations that enable one to predict what will float and what will sink.

Consider another example.[17] Newton discovered a way to describe and predict the motions of two bodies before and after impact using his principle of action and reaction: During impact, two bodies exert a force on each other equal in magnitude but opposite in direction. This principle, in turn, makes reference to a theoretical concept, force, which is defined as mass times acceleration ($F = ma$). If we interpret force realistically, we will think of it as a push or pull of some type.

---

17. Cf. Rom Harre, *Matter and Method* (Atascadero, Calif.: Ridgeview, 1964), pp. 7–18.

We can liken it to what a person feels when he pushes or bends something. (For example, when two billiard balls collide there is a real push [a force] exerted by one on the other. If the ball were conscious, it would feel the force.) However, *force* can be eliminated altogether and no loss of empirical content will occur; we will be able to make the very same observations and predictions whether or not we take the force of impact to be a real entity. *Force* can be eliminated as follows:

$$F = -F'$$

$$ma = -m'a'$$

$$\frac{m(v_i - v_f)}{t} = \frac{-m'(v_1' - v_f')}{t}$$

$$mv_i - mv_f = -m'v_1' + m'v_f'$$

$$mv_i + m'v_1' = mv_f + m'v_f'$$

Here we have eliminated the concept of force altogether. In its place we have introduced the concept of momentum, which enables us to decribe the motion of impact in terms of measurable properties. Some antirealists argue that all theoretical terms can be eliminated in the way that *specific gravity* and *force* have been above. Thus, we should not interpret them realistically as referring to real entities in the world.[18]

Empirically Equivalent Theories

Rational realists make at least two claims, an ontological one and an epistemological one. The ontological claim is that the theoretical entities referred to by the theory exist and the theory describes those entities in a true (or approximately true) way. The epistemological claim is that science is objectively rational in such a way that it is possible in principle to have good reasons for thinking one of a pair of rival theories to be more approximately true than the other.

But the history of science shows a large number of theories that have been empirically equivalent for a long time (Copernicus versus Ptolemy) or that are empirically equivalent in principle (relative or relational theories of space versus absolutist views of space, alternative geometries of space); that is, two or more rival theories entail all the same observational consequences. In this case the theories are

18. For more instrumentalist criticisms of theoretical entities, see Lambert and Brittan, *Introduction*, pp. 150–58.

undetermined by the data; data cannot settle the issue between the rivals in question. In these cases, the rational realist cannot satisfy both his ontological and his epistemological assumptions. If he agrees that one of the rivals is true and the other false, then he must admit that there is no way to tell which is which. On the other hand, if he denies that one is true and one false, he gives up realism altogether.

In cases like these, the realist claims that one can appeal to other things like simplicity, scope of explanation, and the other epistemic virtues mentioned earlier. But antirealists respond in two ways. First, they assert that such a claim is question begging, for according to them the presence in a theory of various epistemic virtues merely shows that the theory works or is successful for our various purposes, not that it is true. Second, they point to the facts that there are several different epistemic virtues a theory may possess, and that usually, given two rival theories, one will be better with respect to some particular epistemic virtue (e.g., simplicity), while the other is superior with respect to another virtue (e.g., internal clarity). How are we to compare conflicting epistemic virtues or conflicting interpretations of the same epistemic virtue (e.g., Cartesian versus Polar coordinates)? If no successful answer to these questions is forthcoming, and none has been given so far, then rational realism is an inadequate philosophy of science.[19]

Successful Theories Can Turn Out to Be False

One of the strongest antirealist arguments points to the success of some false theories. Rational realism holds that a theory will be successful if and only if it is true or approximately true. A theory is successful if it solves a problem, or range of problems, placed before it by embodying several epistemic virtues—in other words, if it explains a set of phenomena (i.e., removes our puzzlement about that set) by offering an explanation that is simple, empirically accurate, predictively successful, internally clear, and so on. If a theory is successful in this sense, then it must be approximately true. Its success is due to its approximate truth.

The history of science, however, presents a surprisingly large number of very successful theories that later turned out to be false (on a

19. For different views about the significance of empirically equivalent theories, see Paul Horwich, "How to Choose Between Empirically Indistinguishable Theories," *The Journal of Philosophy* 79 (1982): 61–77; Hans Reichenbach, *The Philosophy of Space and Time* (New York: Dover, 1958), pp. 10–36; W. V. O. Quine, "On Empirically Equivalent Systems of the World," *Erkenntnis* 9 (1975): 313–28; Newton-Smith, *Rationality of Science*, pp. 19–43.

rational realist interpretation of current theories). These theories explained facts, guided new research, set up and passed various tests, accurately predicted new data, and so on, for a long time—sometimes for two or three centuries. They were successful, but given our current theories, we now believe them to have been false. They were not even approximately true, and in many cases they did not refer to anything real at all. The entities or mechanisms of the theories—aethers, subtle fluids, affinity forces, phlogiston, and the like—are often, like unicorns, nonexistent. So the terms allegedly referring to them were, in actuality, nonreferring terms like the term *unicorn*, or the entities were so poorly described that the theories presented were largely false pictures of them. These theories have not been refined and improved upon by our current theories; they have been abandoned and replaced altogether.

These theories present serious problems for rational realists. How is a realist to explain their success? According to realism, success is due to approximate truth, but these theories were very successful and false. This casts considerable doubt on the truthfulness of our present successful theories. If past theories were successful but later turned out to be false, why should we believe that our current successful theories will not suffer the same fate? The answer cannot be because they are currently successful. So were the past theories. It seems best, therefore, to interpret all scientific theories, past and present, as successful theories that work but are not necessarily true or approximately true.

Larry Laudan has offered a list of past successful theories that later turned out to be false (on a realist interpretation of current theories):[20] Ptolemaic astronomy, chemical affinity theory, subtle fluids chemistry and physics, various aether theories (electrical, caloric, optical, gravitational, and so on), Newtonian mechanics, classical thermodynamics, wave optics, humoral theory of medicine, catastrophist geology committed to a universal deluge, phlogiston theory of chemistry, caloric theories of heat, vibratory theory of heat, vital force theories of physiology, the theory of circular inertia, theories of spontaneous generation, all geological theories prior to the 1960s (which denied lateral motion to the continents), and chemical theories in the 1920s that assumed a structurally homogeneous nucleus. Laudan's list could be extended considerably. According to him, these theories are now regarded as false, but they were very successful, often for a long time. They explained facts, made accurate predictions, and so on.

For example, Augustin Jean Fresnel (1788–1827) used optical aether

---

20. Laudan, *Science and Values*, pp. 7, 110–14, 117, 121, 123, 126, 127.

theory to predict a novel test result: a small beam of light directed at a circular disk will undergo diffraction (bending around the edges of the object) in such a way that a bright spot at the center of the shadow of a circular disc should appear. When tested, the prediction proved correct. But we no longer believe in optical aether theory.[21]

Examples like the ones listed have an important implication. When we express doubt about scientific theories that are currently well established, say by admitting that the science of 150 years from now may show our current theories false, we are not merely engaging in intellectual humility or expressing a scientific epistemological value of considered skepticism or tentativeness. We are expressing the fact that a historical investigation of science should cause us to beware of leaping to the conclusion that our current successful theories are approximately true. Since I am a chastened and eclectic rational realist, I would agree that they may be approximately true, but such claims have been defeated frequently in the past. So when a Christian theist expresses doubt and skepticism about some current theory in conflict with a theological claim, she may be acting in a very rational way. Such judgments should be made on a case-by-case basis.

In sum, the presence of theories like those listed at least weakens rational realism. Some would argue that it justifies abandoning it altogether. James Clerk Maxwell (1831–1879) claimed that aether theories were better confirmed than any other theoretical entity in the science of his day, and such theories have been abandoned. So success is no guarantee of truth or approximate truth or of the existence of the entities and processes postulated in successful theories.

Approximately True Theories Were
Sometimes Unsuccessful

Just as successful theories sometimes turned out to be false, theories now believed to be approximately true or to refer to real entities and processes in the world were often highly unsuccessful for a long time in solving problems and embodying epistemic virtues. Often they were less successful than rival theories that have now been abandoned. So the truth of a theory is no guarantee that it will be successful for some time, and it could easily turn out that the features of a theory that are true do not lend themselves to good scientific testing, and the false features may, in fact, pass several tests, as we have just seen. So an-

21. Cf. P. M. Harman, *Energy, Force, and Matter: The Conceptual Development of Nineteenth-Century Physics* (Cambridge: Cambridge University Press, 1982), pp. 21–24, 112–16.

tirealists argue that the truth of a theory may not cause it to be successful.

Laudan lists the following examples of theories of this type:[22] chemical atomic theory of the eighteenth century, Proutian atomic theory, Wegenerian continental drift theory, wave theories of light before 1820, kinetic theories of heat in the seventeenth and eighteenth centuries, and developmental theories of embryology before the late nineteenth century. Let us look briefly at one of these.

In 1815 the English physician William Prout (1785–1850) drew attention to the fact that if the atomic weights of the chemical elements were computed based on taking hydrogen as one, then a large number of other elements were whole-number multiples of hydrogen or nearly so.[23] Prout hypothesized that hydrogen was the basic constituent of matter—an atom—and that other elements were composed of different amounts of this basic atom. In doing this, Prout led the way toward an atomic interpretation of the chemical elements and unified them by treating them as different combinations of a fundamental particle.

Our current atomic theories agree with the substance of Prout's hypothesis of atomism and, thus, a rational realist would see it as approximately true. But it was not quickly accepted and did not fare well when tested. For one thing, several elements did not empirically confirm Prout's hypothesis but disconfirmed it (Proutians tended to discount this evidence as experimental error). Several elements had atomic weights that were not whole number multiples of hydrogen. Chloride, in particular, had an atomic weight of 35.5, a value at half an integer.

Second, other approaches to chemical phenomena were doing adequate jobs of explaining phenomena, so Prout's theory had fairly successful rivals. Third, when Prout's theories were debated in the late 1800s, biological evolution was becoming popular and was providing a conceptual model for phenomena outside of biology. Specifically, some scientists raised a conceptual problem for Prout's hypothesis: if the elements were related to one another in some way, then perhaps they had evolved from one another over time and there would be no reason to expect their evolutionary pathway to jump in multiples of the atomic weight of hydrogen.

A number of other problems were raised against Prout's hypothesis. By 1860 it was still considered by some scientists "a pure illusion." By 1886 the question of the elements was still unresolved. Not until

---

22. Laudan, *Science and Values*, pp. 110–13.

23. Cf. Maurice P. Crosland, ed., *The Science of Matter: A Historical Survey* (Baltimore: Penguin, 1971), pp. 269–79.

1911–1913 was Prout's hypothesis partially vindicated. Frederick Soddy (1877–1956) showed that different samples (isotopes) of the same element could be isolated with different weights, and that some of the experimental disconfirmation of Prout's hypothesis was due to these isotopes.

The important point of this incident in the history of science, say antirealists, is that while Prout's hypothesis now is considered approximately true by rational realists, it was, for close to a century, quite unsuccessful in competing with its rivals for epistemic virtues. Thus, the approximate truth of a theory is no guarantee of its success, and its lack of success can extend for some time.[24]

An application to creation science may be taken from the history of theories like Prout's. Suppose someone argued, "You have not shown that rational realism is false, you have only shown that it may take a long time for the truth of a scientific theory to show itself. Given enough time, approximately true theories will eventually exhibit good epistemic virtues, even if they do not currently do so." Apart from the problem of defining "enough time," and apart from the fact that an antirealist could respond by saying that this is a question-begging claim against his position, if we grant this rational realist argument, we have the makings of a case for suspending judgment about current inadequacies in creation science. Could it not be that current models of creation science (and they do have many epistemic virtues) show inadequacies not because they are false but because they need more time to develop and be tested?

One could, of course, respond by saying that creation science has had hundreds of years to make its case. But such a response would be wide of the mark. Atomism had been around for two thousand years prior to Prout, and one could have made the same charge against his hypothesis. But Prout's hypothesis had some new features not present in earlier versions of atomism. Similarly, current creation science has new features not present in earlier versions of creationism. So even if, for the sake of argument, we agree that creation science does not fare well compared with its current rivals—neo-Darwinism and punctuated equilibrium theory—the presence of hypotheses like Prout's should give us pause before we discourage the development and test-

---

24. For an analysis of the case of Dalton and atomism, see ibid., pp. 197–207; Harman, *Energy, Force, and Matter*, pp. 120–27; Alan J. Rocke, "Atoms and Equivalents: The Early Development of the Chemical Atomic Theory," in *Historical Studies in the Physical Sciences, Volume 9*, ed. Russell McCormmach and Lewis Pyenson (Baltimore: Johns Hopkins, 1978), pp. 225–63. A brief treatment of the acceptance of the continental drift theory of Alfred Wegener (1880–1930) can be found in I. Bernard Cohen, *Revolution in Science* (Cambridge, Mass.: Belknap, 1985), pp. 446–66.

ing of creation science. It may just need time to develop conceptually in light of modern developments in various areas of science.

Problems with the Cumulative Nature
of Scientific Progress

Rational realists believe that we should accept the referential aspects or approximate truth of current, well-established scientific theories in mature areas of science. One of the reasons for this belief is that, according to them, the history of science has been a development of theories through time in which progress has been largely cumulative. That is, later theories about some domain of investigation usually do not refute and altogether replace their predecessors. Rather, they usually refine them, improve upon them, or preserve them as limiting cases (e.g., Newton's views about the invariance of mass, length, and time are still true in relativity theory if we limit their application to cases where the velocity is much less than the speed of light). Thus, the development of science is a history of progress: theories are getting truer and truer and earlier theories usually are retained and refined by later theories. Thus, we can expect that our current, well-established theories will be preserved and refined by future theories because they are approximately true.

Now suppose one had a different reading of the history of science: The history of science reveals that later theories refute and discard earlier theories. Old theories are not refined but replaced altogether. In fact, later theories give such different pictures of the world from those contained in earlier theories—think of creationist biology in the 1800s compared with current evolutionary theory, phlogiston chemistry versus oxygen theory, quantum physics versus classical physics— that later theories are not even talking about the same things as earlier theories, even though they may use the same words. For example, does Einstein's concept of mass really have anything to do with mass in Newton's theories?

On this reading of science, we do not progress toward truer and truer theories of the world. Rather, the history of science is a somewhat jerky story of replaced theories that worked for a while but then dropped off the scene. If this is true, then why should we have any confidence in the approximate truth of our current theories or in the existence of the things they postulate? A pessimistic induction from the history of science justifies our believing that since most, perhaps all, of past theories were later abandoned, our current theories will be abandoned as well. If this is so, one could argue, then why should the Christian theist worry if a current scientific claim conflicts with what he believes is a well-established theological truth? All he needs to do is wait and the current theory will be abandoned in light of a new one.

It should be clear from this discussion that the nature of progress in science is central to the realist/antirealist debate and plays a role in devising models for integrating theology and science. Let us look at some of the difficulties in the cumulative refinement view of scientific progress that have driven a number of thinkers to a noncumulative replacement view.

THE MEANING AND REFERENCE OF SCIENTIFIC TERMS

If two theories T_1 and T_2, say Dalton's and Bohr's models of the atom, relate in such a way that they succeed in referring to entities that really exist, T_1 is approximately true, and T_2 refines T_1 by being closer to the truth, then both theories must refer to the same thing, atoms in this case, and T_2 does a better job of describing atoms than does T_1.

If the two theories refer to different things, then they are incommensurable, that is, incapable even of being compared with each other. Words can be used equivocally: *red* can mean a color or a communist. If Dalton used *atom* to refer to something so utterly different from what Bohr meant by *atom*, then the shift from Dalton to Bohr simply changes subjects, even though they use the same term to talk about different things. And when a later theory changes the subject of discussion relative to a prior theory, then the two are not comparable and refinement is not possible. Does it make sense to ask whether Darwin's theory of natural selection was an improvement over Newton's laws of motion? Of course not. The two theories don't even discuss the same thing. This would be true even if Darwin had chosen to use the term *force* instead of *natural selection*. It still would not make sense to ask if Darwin's view of force were an improvement over Newton's; the two theories would equivocate on the term *force* and so would be incommensurable.

A number of thinkers have argued that most, perhaps all, cases of theory change in science involve such a radical shift in the meaning of observational and theoretical terms that radical meaning variance occurs, that the two theories concern different things even if they use the same words, and that former theories are not refined but abandoned entirely in such theory changes. Since all older theories have been replaced, we believe that they were false and did not really refer to anything in the world. Therefore, current theories should be read in the same way, and antirealism is established.

The main issues in this antirealist argument involve questions about how terms in a theory get their meaning and refer to things in the world. As David Papineau puts it, "How sentences in scientific theories are to be evaluated as representations of reality depends on the way that scientific terms acquire meaning."[25] How do terms in any language, including scientific discourse, get meaning? How do terms succeed in referring to something in the world? There has been a classic, traditional answer to these questions that, if true, has devastating implications for rational realism. Let us look at the three main ideas involved in the traditional theory of reference and meaning.[26]

First, some believe that the meaning of a general term like *water* or a proper name like *Aristotle* is a set of properties or concepts associated with the term in question. For example, the meaning of *water* can be given by listing some of the key properties associated with water: a clear, odorless, tasteless liquid that freezes into ice, fills lakes, and so forth. The meaning of *Aristotle* is a person who lived in ancient Greece, was the teacher of Alexander the Great, the pupil of Plato, the author of *Nicomachean Ethics*, and so forth. In general, the meaning of a term can be given by listing the set of relevant properties associated with the thing to which it refers.

Second, some believe that the meaning (also called connotation or intension) of a term determines its reference (also called denotation or extension). That is, a term refers to things *through* its meaning. We fix the meaning of a term, and the extension of that term—the things set apart and referred to by it—are those items that satisfy the term, those things that have the properties involved in the meaning of the term. *Water* refers to all the stuff in the world that satisfies the meaning of *water:* stuff that is a clear, odorless, tasteless liquid, and so forth. *Aristotle* refers to that person who was the ancient Greek teacher of Alexander, the pupil of Plato, and so on.

Third, some believe that statements that assert that some item has a certain property are analytic truths, that is, true by definition. "All Ts are P" (e.g., "All examples of water are clear liquids") is true by virtue of the meaning of *water.* Part of the meaning of *water* is "a liquid that is clear." If we later change our view of water (the stuff, not the term), say by agreeing that water need not be a clear liquid, then we change the meaning of the term *water* by changing a central item in the list of properties that constitute the meaning of the term.

25. David Papineau, *Theory and Meaning* (Oxford: Clarendon, 1979), p. 1.
26. For helpful treatments of meaning and reference, see Stephen P. Schwartz, ed., *Naming, Necessity, and Natural Kinds* (Ithaca, N.Y.: Cornell University Press, 1977) and Saul Kripke, *Naming and Necessity* (Cambridge: Harvard University Press, 1972).

We can clarify the traditional view further by applying it to an example. Suppose at a dinner party you see a man sitting at a table drinking from a paper cup, what you believe is water but is actually wine. Suppose further that he is the president of a local bank. Standing next to him is a car salesman drinking, from a wine glass, what you think is wine but is water. You reason that people drink water, not wine, from a paper cup, and wine, not water, from a wine glass. You look at the man who is sitting and say to your friend, "The man drinking water is a local banker." Is your statement true or false?

The answer depends, of course, on which man you are referring to. According to the traditional view of meaning and reference, your statement is false. Why? Because reference is accomplished by whatever satisfies the meaning of the terms you use in referring. You refer to whatever has the list of properties associated with water (i.e., whatever is a clear, odorless, tasteless liquid). The banker sitting down does not satisfy this description, but the car salesman standing by the banker does, so you refer to the car salesman and your statement is false. If neither man was drinking water, then you failed to refer to anything at all. Nothing existed that was drinking water, so you did not refer to anything, even though you thought you did.

Now suppose we supplement the traditional view of meaning and reference with what has been called a holist theory of the meaning of observational and theoretical terms in a scientific theory. The holist theory does not tell us what meaning is. Rather, it tells us how terms get that meaning. According to this view, all terms get their meaning by virtue of the role they play in the entire theory in which they are embedded. The term *electron* refers to a set of properties associated with electrons. Which properties are relevant? All the ones currently embraced in our theories about an electron. *Electron* refers to anything that has properties P_1 through P_n, and these properties define the role *electron* plays in electron theory, namely, "whatever has negative charge and such-and-such rest mass, leaves tracks through cloud chambers, has certain wave properties," and the like.

If the preceding theory of meaning and reference is correct, then an important conclusion follows: Whenever a theory T_2 changes one of the roles of a term that was also in T_1, the new theory does not mean the same thing by using that term; it equivocates such that there is radical meaning variance regarding the term in question. Therefore, the new theory replaces the old one instead of refining it, and the old theory did not refer to anything at all because, in light of current theory, we no longer believe that there ever was anything in the world that satisfies the properties associated with the role that term played in the discarded theory.

An example may help clarify the point. Consider different theories about the electron held by J. J. Thomson, N. Bohr, and present physical theory.

Thomson: Electrons$_t$ are $T_1, T_2,.., T_n$.
Bohr: Electrons$_b$ are $B_1, B_2,..., B_n$.
Ourselves: Electrons$_o$ are $O_1, O_2,..., O_n$.

Thomson referred to whatever there was in the world that satisfied all of the properties T_1 through T_n involved in his theory of the electron; likewise for Bohr and our current theories. Now the transition from Thomson to Bohr to us has shown that, in light of current theory, nothing ever existed that had all of the properties involved with the role electrons played in the theories of Bohr and Thomson. Electrons for Thomson were nonorbiting, negatively charged particles embedded in an atom, much as raisins are embedded in plum pudding, and they were held together by a sphere of positive charge surrounding them. The entire mass of the atom was due to electrons (Thomson thought that there were 1836 electrons in an atom of hydrogen). In addition, electrons were conceived of as fundamental particles, they had Newtonian mass invariant with respect to velocity, they traveled through absolute space and time, they could not be converted into energy, and they were not waves.

It may be thought that at least Thomson was right about an electron having negative charge. But remember, the holist theory of meaning holds that the meaning of a term is determined by *all* of the properties associated with it, not just one or two, and if any change, the meaning of the term changes.

Furthermore, the theoretical notion of charge has itself changed through time. Some scientists used to think that charge was a Newtonian gravitational force, others that it was an electrical, ether fluid that constituted the affinity between matter (the fluid had repelling, elastic properties and it was "subtle," i.e., it could penetrate the empty spaces between ordinary matter), others that it was an electrical "sphere of activity," an empty space wherein electrical forces manifested themselves, still others that it was a filament tube or line of force formed as vortices in the swirling etherial fluids between matter, and so on.

Thomson himself was uncommitted on the question, but he seems to have favored the view that forces, electrical ones included, were to be understood in terms of the ether viewed as some sort of fluid, a solution that presupposes absolute space in which the fluid rotates.[27] The point here is simply that even notions like electric charge and force of attraction have undergone significant change and, some antirealists would argue, radical meaning variance. So when two theories appear to agree about some term (*electron* refers to particles that have charge for Thomson, just as it does for us), the agreement may be superficial only.

Bohr's model differs from Thomson's. For Bohr, electrons orbit, and they do so only in specific orbitals. Our current model differs from both. If the traditional theory of meaning and reference and the holist theory of meaning for observational and theoretical terms is correct, then there never was anything in the world referred to by *electron*$_t$ or *electron*$_b$. Acceptance of either theory did not imply that anything answering to the theory existed in the real world, and these theories were false. By extension, our present theories will change, so they do not refer to anything real, so they are not approximately true either. So the history of science should be read as a series of discontinuous replacements, even when later theories mask this fact by using the same words, instead of as a continuous refinement of earlier theories converging on truer and truer pictures of the world.

Rational realists have responded to this challenge in at least two ways. First, they have offered different theories of meaning and reference. The issues involved in these responses, and counterarguments by antirealists, are too complex to cover here. I have surveyed some of them elsewhere.[28] Suffice it to say that if the classical theory of meaning and reference and the holist theory of meaning for observation and theory terms is correct, then rational realism is in trouble. And whether rational realist responses are correct or not, the debate about meaning and reference itself illustrates the dependence of science upon philosophical presuppositions—meaning and reference, in this case— and, thus, scientism is refuted.

A second response by rational realists is to say that a term does not get its meaning from its entire role in a theory, but only from some specific subset of roles. We do not need to agree with Thomson about all the alleged properties of an electron to be talking about the same thing; we only need to agree with some small set of properties or some

27. See Harman, *Energy, Force, and Matter*, pp. 79–84.
28. See J. P. Moreland, "The Scientific Realism Debate and the Role of Philosophy in Integration," *Bulletin of the Evangelical Philosophical Society* 10 (1987): 38–49. Two different realist responses to antirealist theories of reference and meaning are offered by Hartry Field in "Theory Change and the Indeterminacy of Reference," *The Journal of Philosophy* 70 (1973): 462–81, and Newton-Smith, *Rationality of Science*, pp. 148–82. Papineau offers antirealist responses to both realist strategies in *Theory and Meaning*.

small set of indicated phenomena to be explained (various electrical phenomena like electrical batteries, charges in wires, magnetic fields, and the like).

As an eclectic realist, I sympathize with this objection but urge caution. Even if there are several cases where our current theories refine previous ones instead of replacing them, we are likely, for two reasons, to interpret the number of such instances as larger than it is. First, our present bias in favor of realism inclines us to think that earlier theories meant what we now mean when they used the same terms (*mass, electron, evolutionary development*) we use today, giving a superficial appearance of agreement.

Second, we lack familiarity with the details of the history of science, including the various aspects of the theories in question. There often are many more differences than similarities between past and present theories. This makes it more difficult to specify what and how many features of the compared theories must be held in common before we count them as cases of refinement and not replacement. Such decisions can be arbitrary. When we recall that everyone agrees that many past theories have simply been replaced and abandoned (e.g., no one refers to phlogiston in chemistry any longer; see also the theories listed that were successful but later turned out to be false), then even if one is a rational realist, she should not be too dogmatic about her position.

THEORIES ARE EPISTEMICALLY INCOMMENSURABLE

We have just seen that antirealists argue against theory refinement because, allegedly, theories are semantically incommensurate. They cannot be compared because they equivocate on the meaning of the terms they employ. Others have argued that theories are epistemically incommensurable. That is, different theories have their own standards of rational appraisal, no theory-neutral considerations can be used to judge between two theories, and, thus, later theories cannot be compared with earlier ones to determine which are more rational.

Thomas Kuhn has argued that different scientific theories have different epistemic virtues, there is no way to ajudicate disputes about which epistemic virtue is better, and the only standards for evaluation of a particular theory are those available from within it. For example, two rival theories may differ in that one appears better when judged by certain epistemic virtues (simplicity, internal clarity, and limited external problems) while the other appears better when judged by other epistemic virtues (empirical accuracy, predictive success, fruitfulness for guiding new research). In such cases, how can we judge between the theories? Adherents of the first theory will argue that simplicity and the like are more important; adherents of the second theory will argue that empirical accuracy and the like are more important. In these cases, there is no way to tell which is better, so later theories cannot be considered improvements or refinements of earlier theories; they can only be considered different theories.

It has also been argued that two theories will differ about other epistemological issues, for example, the appropriate methods to use, the proper instruments to be employed, the problems and questions considered important to answer.[29] For example, Aristotle's view of motion involved solving the question *Why does something move?* Rest was considered natural and not in need of explanation, but motion was considered unnatural and in need of explanation. Isaac Newton's (1642–1727) view of motion reversed this: uniform motion was considered natural and coming to rest involved the application of forces. Thus, the questions, conceptual apparatus, methods, and instruments involved in two theories make rational appraisal between them impossible. Later theories do not refine and improve earlier ones as far as truth is concerned; they are merely different replacements of those theories.

We will see in the next chapter that an all-out agreement with epistemic incommensurability commits one to conceptual relativism. If there are no neutral criteria for judging between theories, then there is no way to apply judgments that claim that one theory is truer or more rational than another. But one does not need to agree that two theories cannot be compared to appreciate the point made by antirealists. For theories often do differ over the methods they employ, the problems they try to solve, and the standards of rationality they appeal to. In cases like these, it may be very difficult to claim that one is better than another. When this is the case (when current theories are compared with earlier ones), we may be wise to say that current theories work better than earlier ones *given our present purposes.* But it would be unwise to claim that present theories are truer or more rational than earlier ones. Only a case-by-case examination of the theories themselves can decide the issue.

It seems to me that the creation/evolution controversy needs to be viewed in light of differing epistemic issues as well as in light of empirical data. For this dispute involves, among other things, different views of the nature of science, of the source and role of external conceptual problems, of the relative weights of empirical anomalies (gaps in the fossil record, problems with prebiotic soup experiments), of

---

29. Kuhn, *Structure of Scientific Revolutions*, 2d ed., enlarged (Chicago: University of Chicago Press, 1970), especially chaps. 8–10; Rescher, *Limits of Science*, pp. 35–111.

proper methodology (e.g., some creationists charge that in origin-of-life experiments the scientist interferes to facilitate reactions in ways that nature herself could not accomplish), of the importance of prediction, and so on. Disputants in this controversy seem to be talking past each other, and the source of some of this lack of communication can be found in these epistemological problems.[30]

Realism and Antirealism Are Empirically Equivalent Second-Order Theories about Science Itself

Realists often claim that one cannot explain the success, the progress, the growth of science without assuming that rational realism is true. Without a rational realist interpretation of science, the success of science would be an unexplained miracle, for only if we really do succeed in referring to entities in the world and increasing the accuracy of our descriptions of those entities can we explain why scientific practices are so successful in solving problems.

Antirealists dispute this claim as question begging. They claim to be able to describe the actual history of the development of science as well as, and perhaps better than, rational realists. That is, they claim that rational realism and antirealism are empirically equivalent theories about the nature of science. Antirealist arguments for this claim usually boil down to two main lines of defense. First, antirealists claim to be able to spell out a theory of scientific rationality that makes sense of science but does not appeal to truth as correspondence or successful reference to unseen theoretical entities. Like life forms in the evolutionary struggle for survival and reproductive advantage,

---

30. Two other issues are often discussed in connection with the replacement/refinement debate. First, there is a debate about what should be the primary units of scientific change or growth, e.g., fairly low-level theoretical generalizations like Dalton's model of the atom versus Bohr's model, or broader research programs, e.g., atomism itself. If low-level theories are in view, a refinement interpretation of the history of science is harder to sustain. If the basic unit of science is the research program, then more continuity can be sustained. Unfortunately, there are debates about what holds the several theories together in a research program and makes them *one* research program (e.g., all of the various theories of the atom). Is it a shared group of questions, a shared group of theoretical entities believed to exist, or what? Also, the broader the research program is, the harder it is to claim that it is easily falsifiable and the more difficult it is to show the superiority of one research program over a rival. For more on this, see Laudan, *Progress and its Problems*, pp. 70–120; Imre Lakatos, "Falsification and the Methodology of Scientific Research Programmes," in *Criticism and the Growth of Knowledge*, ed. Imre Lakatos and Alan Musgrave (Cambridge: Cambridge University Press, 1970), pp. 91–196. Second, there is a debate about what it means to say that a later theory refines an earlier one. See Laudan, *Science and Values*, pp. 124–36.

theories are in a struggle for allegiance. Theories that cannot compete—that do not appropriately embody good epistemic virtues—simply die off and are discarded.[31] In the next chapter we will look more carefully at the details of antirealist accounts of scientific rationality that avoid realist assumptions. If such accounts are themselves successful, realist appeals to the success of science are neutralized.

Second, antirealists claim that the history of science is on their side. As we have seen in this chapter, several past theories worked well for long periods of time and later were falsified, and several theories that are approximately true in light of realist interpretations of current theories were highly unsuccessful for a long time. Antirealists claim that these theories are hard for the realist to explain, for according to realism, the success of a theory is tied to its approximate truth. Thus, these theories are anomalies for scientific realism but are expected by an antirealist interpretation of science.

More work needs to be done in the philosophy and history of science, and more Christians need to work in these important fields. Suffice it to say that a major debate rages today among intellectuals about how to understand science and its history. The disputants are scholars who are fully appraised of the issues involved in interpreting science. Some of them see science as a progressive, cumulative enterprise where later theories (usually) refine and preserve the good features of earlier ones as science advances in its quest for truer and truer descriptions of the world. Others see science as progressively solving problems by (usually) refuting earlier theories and replacing them with later ones. Even if the truth lies somewhere between these extremes, one should be cautious in dogmatically claiming that current scientific theories are true or approximately true. They may be, but they might be found false and replaced in the future instead. Thus, we have another reason to understand the theology/science conversation as a dialogue and not a monologue.

Summary and Conclusion

Rational realists have felt the force of antirealist criticism. For example, Ernan McMullin, one of the leading scientific realists, makes this claim:

> The basic claim made by scientific realism, once again, is that the long-term success of a scientific theory gives reason to believe that something

31. Cf. Bas C. van Fraassen, *The Scientific Image* (Oxford: Oxford University Press, 1980), pp. 34–40.

like the entities and structure postulated by the theory actually exists. There are four important qualifications built into this: (1) the theory must be successful over a significant period of time; (2) the explanatory success of the theory gives some reason, though not a conclusive warrant, to believe it; (3) what is believed is that the theoretical structures are *something like* the structure of the real world; (4) no claim is made for a special, more basic, privileged, form of existence for the postulated entities. These qualifications: "significant period," "some reason," "something like," sound very vague, of course, and vagueness is a challenge to the philosopher. Can they not be made more precise? I am not sure that they can; efforts to strengthen the thesis of scientific realism have, as I have shown, left it open to easy refutation.[32]

Some of the factors examined in this chapter have caused realists like McMullin to be vague and tentative in their realism. In light of these factors, we can draw four conclusions.

First, once again we see how philosophy is presuppositional to science. The validity of scientific claims about the world rests, in part, on the truth of rational realism. If antirealism is true, then this would count against a scientist who claims that some theory is approximately true, or that some theoretical entity really exists because his theory is successful. Even when rational realists try to use science to justify itself, they justify this very appeal by offering philosophical arguments for such an inference to the best explanation (i.e., "the best explanation for the success of science is rational realism") and by offering philosophical arguments against antirealists. In light of this, scientists and theologians should become more familiar with the philosophical (and historical) aspects of science.

Second, even if one remains an extensive rational realist, that is, even if one embraces rational realism as an interpretation of all areas of science—biology, psychology, physics, and their subfields—he should do so in a nondogmatic way. The debate about realism is simply too complex, it involves too many issues, and it has too many good thinkers on both sides to be ignored and dismissed as philosophical pedantry. Thus, when one uses a scientific argument for or against theology, he should be cautious.

Third, the best approach may be an eclectic one embracing rational realism as an interpretation of some particular theory and antirealism as an interpretation of some other theory. I will try to offer some illustrations of this approach in the next chapter, after we have looked at some alternatives to realism. But if this approach is correct, then

32. Ernan McMullin, "A Case for Scientific Realism," in *Scientific Realism*, ed. Jarrett Leplin (Berkeley: University of California Press, 1985), p. 26.

it would be appropriate to adopt realism or antirealism on a case-by-case basis, depending on which approach does better justice to all the considerations relevant to the case at hand.

Finally, the realist/antirealist debate complicates the interaction between theology and science. If some scientific area seems to conflict with a theological claim, then a solution to integrating these areas should not automatically assume the approximate truth of the scientific side of the discussion, even if that scientific claim has been successful. It may be that one should understand that area of science in antirealist terms. Or it may be that one should interpret it in realist terms and try to adjust his theology with appropriate respect for biblical inerrancy and correct exegesis. But in cases like these, there may be conflicts between science and theology, and by adopting a tentative realist interpretation of the scientific issue one can still claim rationality in suspending judgment about how the two areas will eventually harmonize. For the scientific area may turn out false. In any case, the philosopher should be involved with the scientist and the theologian in the task of integration.

Alternatives to Scientific Realism

Contemporary opponents of the classical Darwinian account of the origin and differentiation of species, for instance, variously claim that evolution is no more than a "theory" (as opposed to a fact), that it is not a *"scientific* theory" (because, it is claimed, evolutionary hypotheses are not in principle falsifiable) and that it is not a *"well-confirmed* theory" (because, so "scientific creationists" say, there are too many experimental facts for which it cannot account). These anti-evolutionary claims cannot be understood, much less settled, in the absence of a clear understanding of what theories are, how they relate to facts, and why we choose some in preference to others.

> —Karel Lambert and Gordon Brittan, Jr.,
> *An Introduction to the Philosophy of Science*

Most Christian scientists currently appear to favor a "realist" view of scientific theories. Apparent conflicts between science and Scripture are then generally resolved by modifying either our reading of Scripture or the offending scientific theories. [An antirealist] alternative enables one to make use of the practical results of scientific theories while at the same time withholding any commitment as to the validity of their epistemological content.

> —John Byl, "Instrumentalism: A Third Option"

In the previous chapter we saw that several difficulties present themselves to anyone who embraces scientific realism as a philosophy of science. These difficulties may not warrant a wholesale abandonment of scientific realism, but they do show that a naive acceptance of scientific realism is unjustified. An alternative to scientific realism may be the best road to take, either as an entire philosophy of science or as an approach to be integrated with scientific realism into an eclectic philosophy of science wherein a realist or antirealist stance is

171

taken on a case-by-case basis. As the quotes at the beginning of the chapter illustrate, one's philosophy of science is foundational to how one integrates science and theology.

This chapter focuses on two tasks. First, we will look at some of the major alternatives to scientific realism. Second, we will look briefly at an eclectic approach to science by considering some areas of science where an antirealist stance may be preferable to a realist one.

Varieties of Antirealism

At the beginning of chapter 4, we categorized the different options in the realist/antirealist debate as in figure 5.1.

Rational Nonrealism (Instrumentalism)

Rational nonrealism, also called instrumentalism, is a group of different views about science that are united around two basic themes. First, instrumentalists agree that science is an objectively rational approach to the world. That is, instrumentalists deny a conceptual relativist interpretation of science, namely, that science is no more rational than astrology, that one scientific theory is no more rational than another, and that what is rational to one person or community is not necessarily rational to another. In rejecting conceptual relativism, instrumentalists agree that rationality is an objective, normative notion and that it can be objectively rational to believe or accept science in general or a specific scientific theory in particular.

Second, instrumentalists agree that scientific theories are not true or approximately true descriptions of the theoretical entities, structures, or processes that underlie our observations. Science does not progress toward a true picture of the way the world is. Rather, a scientific theory is a device or instrument that is justified by its utility, not its truthfulness. Scientific theories "work," they are successful in accomplishing the purposes for which they were formulated.

Figure 5.1
Options in the Realist/Antirealist Debate

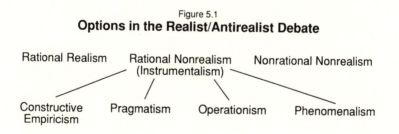

But what are these purposes? What does it mean to say a scientific theory "works?" Different answers to these questions are given by different versions of rational nonrealism.

PHENOMENALISM

Epistemology is the branch of philosophy that focuses on issues in the theory of knowledge: What is knowledge? Can we know anything? What can we know? How do we justify our knowledge claims? Phenomenalism is essentially a radical empiricist theory of epistemology that stems from the idea that all our knowledge is derived from and is about immediate sensory experiences.[1] Only propositions about directly observable phenomena are meaningful or have the status of knowledge, and claims about unobservable theoretical entities are really disguised statements about sensory experiences and must be translated or reduced to such. Furthermore, phenomenalists hold that experience is nothing more than a succession of immediately given, conscious states known by direct inspection or acquaintance with the flow of one's own experiences. These conscious states are variously called ideas, sense data, or impressions.

An example may help to clarify a phenomenalist theory of knowledge. Consider the claim, "There is a pink rose in the next room." On the surface, this seems to be a claim about an entity, a pink rose, that exists and has its properties independently of anyone's observing it. But according to phenomenalists, physical objects are logical constructs out of sense data, where a sense datum is a particular color patch, a particular sensory shape, and so on. Thus, the claim, "There is a pink rose in the next room," is identical to a potentially infinite set of claims like this: "If I were to go into the room at point x_1 and look at point y, I would have a pink, rose-shaped sense datum;" "If I were to go into the room at point x_2 and look at point y, I would have another pink, slightly different rose-shaped sense datum," and so forth. In other words, statements about physical objects that exist are reducible to or translatable without remainder into a set of statements about actual or possible experiences of specific sense data.

As a theory of knowledge, phenomenalism also goes by the name reductionism, positivism, or extreme empiricism, and it has been embraced, in one form or another, by a number of thinkers, including

1. For an analysis of phenomenalism as a general empiricist strategy in epistemology, see A. J. Ayer, *The Problem of Knowledge* (Baltimore: Penguin, 1956), pp. 75–83, 118–29; Michael Williams, *Groundless Belief: An Essay on the Possibilities of Epistemology* (New Haven, Conn.: Yale University Press, 1977). An overview of empiricist theories of meaning can be found in Richard Rorty, *The Linguistic Turn* (Chicago: University of Chicago Press, 1967), pp. 1–124.

Berkeley, Hume, Mill, Russell and Rudolph Carnap (earlier in their careers), Karl Pearson, Ernst Mach, and the mid-nineteenth-century chemist Benjamin Brodie (1783–1862).[2]

As a philosophy of science, phenomenalism implies that science does not give the truth about unobservable theoretical entities (e.g., hydrogen atoms or magnetic fields) that are used to explain what causes observable phenomena. Rather, science merely attempts to "save the phenomena," that is, to facilitate our ability to describe the successive sensory experiences we have of the world and to anticipate future sensory experiences. As Mach put it, "it is the object of science to replace, or *save*, experiences, by the reproduction and anticipation of facts in thought."[3] Phenomenalism as a philosophy of science can be clarified by focusing on four issues: What is it that gives us certain, scientific knowledge? What is the nature of scientific laws and theories? What is a sense datum? And what is the nature of scientific explanation?

First, phenomenalists hold that we only have real knowledge that is cognitively meaningful and scientifically verifiable when that knowledge is about immediate sensory data or facts. These sensory data are the things we know directly by aquaintance with our own mental life, and they provide a permanent foundation for erecting the edifice of scientific knowlege. Science is about what we can see, and any thing or process that cannot be perceived cannot be supposed to exist for science. For example, if a chemical solution changes from a clear liquid to a pink liquid during an acid/base titration experiment, the scientist has a series of sensory experiences of something clear and wet that is here now, followed by something pink and wet that is here moments later. The scientist knows these experiences by merely attending to his mental life during the experiment.

Second, theoretical terms, theories, or scientific laws are merely summaries of data, labor-saving mental devices for classifying observations. A theory is a mnemonic device by which sensory facts are stored and recollected; it is an elliptical formula of the relations of dependence that regularly occur between observable events and properties. Thus, theoretical terms, theories, or scientific laws can be translated without remainder into a set of statements about actual or possible observations. For example, the word *atom* refers not to an entity that exists but to a potentially infinite set of actual or possible

---

2. Cf. Bertrand Russell, "The Relation of Sense-Data to Physics," reprinted in *Philosophy of Science*, ed. Arthur Coleman Danto and Sidney Morgenbesser (New York: Meridian, 1960), pp. 33–54; Ernst Mach, *The Science of Mechanics* (La Salle, Ill.: Open Court, 1960).

3. Mach, *Science of Mechanics*, p. 577.

sensory experiences in the laboratory. A law of nature is merely a description of the succession of sensory phenomena by means of which we can predict or anticipate future experiences; it is not a causal law.

Some examples may help. The statement, "there is presently an electric current in the wire on the table," appears to refer to an unobservable theoretical entity, an electric current. According to phenomenalism, this statement can be completely translated so as to remove reference to this theoretical entity and replace it with a set of statements like this: "If the galvanometer on the desk there were introduced into this circuit, the pointer would be deflected from its present position."

Again, the word *hydrogen* refers not to an atom that exists but to a set of laboratory observations of "hydrogenated space" (i.e., a set of laboratory observations of colorlessness, weight, and volume that are experienced after a series of prior sensory experiences occur). These prior experiences are those associated with the production of hydrogenated space, that is, a series of observations that realists take to be the steps necessary for preparing hydrogen gas. The laws of chemistry are codified, summary records of past experiences that can be used to predict or anticipate future experiences. These laws allow one to predict what observations will occur after certain observed chemical operations take place. "$2H_2 + O_2 = 2H_2O$" means not that two molecules of hydrogen and one molecule of oxygen produce two real entities known as water molecules but that, in the past, two units of a certain observed volume, weight, and odorlessness produced by certain techniques have been added to one unit of volume, weight, and odorlessness produced by certain other techniques, and a wet, colorless substance of two units has always followed.

In chapter 4 we discussed an argument, advanced by Berkeley and Mach, to the effect that Newtonian forces of attraction, repulsion, and collision do not need to be taken as real, hidden entities residing as powers in physical bodies. Rather, they can be treated as mere mathematical summaries that remind us of how the series of observational experiences of moving bodies have proceeded in the past. On this basis, we can predict that two bodies with certain observed weights and velocity will move in certain ways after their surfaces are observed to be in contact.[4]

Third, the ultimate entities to which theoretical entities are reduced are sense data or impressions. As Ernest Nagel put it, "The psychologically primitive and indubitable objects of knowledge are the im-

---

4. For an overview of this type of argument, see R. Harre, *Matter and Method* (Atascadero, Calif.: Ridgeview, 1970), pp. 8–18.

mediate 'impressions' or 'sense contents' of introspective and sensory experience."[5] Bertrand Russell remarked, "When I speak of a 'sense-datum,' I do not mean the whole of what is given in sense at one time. I mean rather such a part of the whole as might be singled out by attention: particular patches of color, particular noises, and so on."[6] An example of a sense datum would be a particular experience of a certain shade of red, an experience of a certain shape, or of a certain noise. Physical objects in general and theoretical entities in particular are reduced to series of appearances, bundles of particular sense data. Thus, a rose is a bundle of a particular color, a particular shape, a particular texture, and so on. Hydrogen is reduced to a series of certain laboratory experiences that regularly follow other experiences.

Fourth, the role of a scientific theory is not to explain observational phenomena by postulating causal entities and processes responsible for them. Rather, a theory is merely a descriptive summary, given in an economical way, that describes the succession and concomitance of sensory experiences. By viewing theories in this way, phenomenalists avoid the metaphysical postulation of occult (i.e., unobservable or undetectable) entities, processes, and relationships that realists take to underlie and be causally responsible for observable phenomena.

Since the purpose in this chapter is primarily to state alternatives to scientific realism, I do not wish to dwell on criticisms of phenomenalism. It was a dominant philosophy of science up to the middle of this century, but today few professional philosophers of science would classify themselves as phenomenalists. The reasons for the demise of phenomenalism, and of empiricism in general, are many, but some follow.

First, the epistemological principle of phenomenalism, at least that variety of phenomenalism that was popular in the first half of this century, namely, logical positivism, is self-refuting. The principle "The only proposition that we can know or that is cognitively meaningful is one that is empirically verifiable" is not itself empirically verifiable. Thus, it fails to satisfy its own criterion of acceptability.

Second, in spite of protests to the contrary, there is simply a difference between something existing and its being an actual or possible object of perception. For example, it is one thing to say a pink rose exists in the room, but this is not identical to the claim that I can have a potentially infinite set of rose-type sensory experiences. The rose is what causes the experiences we have of it, and a rose, or a magnetic

---

5. Ernest Nagel, *The Structure of Science* (Indianapolis: Hackett, 1979), p. 120. See also pp. 117–40.

6. Russell, "Relation of Sense-Data," p. 35.

field for that matter, can exist without being seen (or, in the case of the magnetic field, without being observable even in principle). Conversely, a person can have rose-type experiences, say in a hallucination, without there actually being a rose in the room. So physical objects in general and theoretical entities in particular exist and have properties, and their existence and nature are not identical to our perceptions.

Third, philosophers who criticize the observation/theory distinction claim that there is no such thing as a brute confrontation with an uninterpreted sense datum that is simply given to experience. Rather, experiences, especially scientific ones, presuppose some interpretive framework before they can be intelligible and significant. Even if there is a meaningful distinction between observation and theory—and I think there is—it is not easy to draw. And the typical kinds of observations that scientists make require a lot of background theory to know what to look for and to understand what is being seen. So the standard phenomenalist appeal to color patches, specific shapes, or specific noises is a long way from the types of observations that often take place in science. When a scientist claims to see an electron leaving a track through a cloud chamber, she is applying a fairly sophisticated interpretive grid to her experiences; she is not merely reading off of her immediate sensory images a series of particular observations.

Because of these and a host of other criticisms that have been leveled against phenomenalism, few today claim to be in this camp. Thus, phenomenalism is not especially helpful in an attempt to find viable antirealist interpretations of science. But before we move on to the next view of science, there is an important lesson to be learned from our investigation of phenomenalism. Among those not familiar with philosophy, including a number of scientists, it is fairly popular today to claim that science is a field that focuses on testable sensory experience. Science, so we are told, is the domain of observation. Theology and other fields do not use observation as science does.

Anyone who tries to press this point too far should study what a full-blooded, consistent empiricist understanding of science looks like. As we have seen, this phenomenalist understanding of science abandons the existence and knowability of a world of three-dimensional, mind-independent objects, it knows nothing of the law of cause and effect (only temporal and spatial succession can be observed), and it abandons science as an explanatory enterprise that uses unobservable, theoretical entities that are postulated as causally responsible for observable phenomena. This is why full-blooded empiricism has always been at odds with scientific realism. But a scientific realist understanding of science is often unknowingly assumed by people who allow the prestige of science (derived from a realist view) to be a vantage point from which to judge as irrational any field that supposedly does not use empirical testing as its calling card.

One can adopt a weaker emphasis on the value of empirical testing (as indeed one should in light of the phenomenalist extremes we have just discussed), and in this sense science usually does use empirical testing. But it is hard to define closely just what role observation plays in science. Surely it is important, but as we saw in chapter 1, it plays different roles in different areas of science; some items of scientific discussion (anxiety states, attractive atomic forces, magnetic fields) are not directly observable in principle but are connected to observation through a web of background theories about the items in question, the equipment being used in the experiments, and the like.

Furthermore, theology and other fields do appeal to observations. The existence of various kinds of design, order, and beauty in the world, the evidence for the resurrection of Jesus, and the changed lives of people who have met him are all observational data used by areas of theology.

So while phenomenalism currently may not be a popular model of science, a study of phenomenalism can pose a warning to anyone who makes a naive appeal to the observational aspects of science, who tries to make observation a line of demarcation between science and nonscience, or who tries to pour the bulk of scientific theorizing and testing into the mold of observation. Such an attempt has already been tried, and those tempted to repeat this view of science should take a serious look at phenomenalism before they begin.

OPERATIONALISM

Operationalism, occasionally called operationism, is an approach to science very similar to phenomenalism. It is a form of positivism (an empiricist approach to epistemology emphasizing the role of empirical testing and verification for the meaning and justification of truth claims) originally associated with the views of Harvard physicist and mathematician Percy William Bridgman (1882–1962).[7]

Whereas phenomenalism links scientific laws and theories to actual or possible sensory experiences, operationalism links them to actual

7. See P. W. Bridgman, *The Logic of Modern Physics* (New York: Macmillan, 1927); *The Nature of Physical Theory* (Princeton, N.J.: Princeton University Press, 1936). For helpful secondary sources, see John Losee, *An Historical Introduction to the Philosophy of Science,* 2d ed. (Oxford: Oxford University Press, 1980), pp. 175–78; Carl G. Hempel, *Philosophy of Natural Science* (Englewood Cliffs, N.J.: Prentice-Hall, 1966), pp. 88–97; Hempel, *The Aspects of Scientific Explanation* (New York: Macmillan, 1965), pp. 123–33; *The Encyclopedia of Philosophy,* s.v. "Operationalism," by G. Schlesinger.

or possible laboratory operations.[8] For the phenomenalist, scientific laws and theoretical terms really refer not to mind-independent entities and events but to mind-dependent sensations. For the operationalist, scientific laws and theoretical terms really refer to experimental activities or operations. Thus, readings on a screen or tracks through a cloud chamber are not to be taken as evidence for the existence of unobservable particles; rather, statements about particles are really disguised statements about the readings or tracks themselves when they are recorded in one's laboratory notebook when certain operations are performed.

Bridgman stated the central tenet of operationalism this way: "In general, we mean by any concept nothing more than a set of operations; *the concept is synonymous with the corresponding set of operations*. . . . The proper definition of a concept is not in terms of its properties but in terms of actual operations. . . ."[9]

Bridgman's favorite illustration of an operational definition involved the physical concept of length. According to operationalism, length is not a property of bodies discovered by means of measurements; rather, a statement that something is three feet long is a disguised statement about the procedures one follows in measuring the thing in question (e.g., placing a ruler end-to-end along the surface of the thing in question). Length is not an attribute of a physical object; it is a set of operations of sliding rulers, marking coincidences, and recording one's operations. Bridgman defined length this way:

> We start with a measuring rod, lay it on the object so that one of its ends coincides with one end of the object, mark on the object the position of the other end of the rod, then move the rod along in a straight line extension of its previous position until the first end coincides with the previous position of the second end, repeat this process as often as we can, and call the length the total number of times the rod was applied.[10]

It is important to be clear about what Bridgman is claiming. The length of a table is not an actual property of the table itself. Rather,

8. Laws and theories were also linked to "pen-and-paper" operations, that is, mathematical calculations in a laboratory notebook. Thus, the stress of a body (external or internal forces on a body that cause deformation) can be operationally defined via Hooke's law (the strain in an elastic body is proportional to its stress) by a mathematical quantity calculated from the strain (the actual alteration or deformation of the body), which, in turn, can be operationally defined in terms of observations.

9. P. W. Bridgman, "The Logic of Modern Physics," in *The Logic of Modern Physics*, reprinted in *The World of Physics*, ed. Jefferson Hane Weaver (New York: Simon and Schuster, 1987), 3: 840.

10. Ibid., p. 842.

the physical concept of length *is identical to* a set of measuring operations. If we used two different procedures to measure the table, then even if they agreed that *six feet* was the result, there would be two completely different notions of *length* applied.

In general, an operational definition is a rule stating that the defined term (*length* in the preceding example) is to apply to a specific case if and only if the performance of specific operations in that case gives a specified characteristic result. Another example would be an operational definition of *harder than*. We normally take this to specify a comparison between the actual hardnesses of two or more materials. But according to operationalism, *harder than* can be defined by this rule: "Material a is called harder than material b just in case the operation of drawing a sharp point of a across the surface of b results in a scratch mark on b."

A scientific term is meaningful if and only if it can be given a unique operational definition. Another example of this would be an operational definition of the "property" of being magnetic. To say that something is magnetic is to say nothing more nor less than that if iron filings are placed close to the thing in question, they will be attracted by the ends of the bar and cling to it. Similar operational definitions can be given to all other theoretical concepts in physics and chemistry such as velocity, temperature, electric charge, acid, and so on.[11]

There are also biological examples of operational definitions. David Hull has pointed out that some biologists defined genes not as real entities in living organisms but as the operations and recording thereof of certain steps in breeding experiments.[12] Another biological example of an operational definition is the concept of *fitness*. A number of thinkers have criticized the Darwinian notion of the survival of the fittest as being circular. What will survive? The fittest. What are the fittest? Those that survive. Some biologists have attempted to avoid this problem by defining *fitness* in terms of the differential perpetuation of genotypes that can in turn be spelled out in operational terms that focus on directly observable manipulations or recordings of data.[13]

11. Cf. Rom Harre, *Philosophies of Science: An Introductory Survey* (Oxford: Oxford University Press, 1972), pp. 78–80.

12. David Hull, *Philosophy of Biological Science* (Englewood Cliffs, N.J.: Prentice-Hall, 1974), pp. 17–19; "The Operational Imperative—Sense and Nonsense in Operationism," *Systematic Zoology* 16 (1968): 438–57.

13. For philosophical discussions of problems in specifying the notion of survival, natural selection, and fitness, see Mary Williams, "The Logical Status of Natural Selection," in *Conceptual Issues in Evolutionary Biology: An Anthology*, ed. Elliot Sober (Cambridge, Mass.: MIT Press, 1984), pp. 83–98; Elliot Sober, *The Nature of Selection: Evolutionary Theory in Philosophical Focus* (Cambridge, Mass.: MIT Press, 1984), pp. 61–85.

Operational definitions have also been employed in psychology, especially in behavioristic approaches to psychology.[14] The more radical forms of behaviorism in psychology hold that psychology is not the study of inner mental events, processes, or entities, but of publicly observable behavior. Thus, the need for a father figure is to be defined in terms not of inner feelings of anxiety or loneliness, but of behavioral tendencies in certain circumstances; for example, the need for a father figure is identical to the tendency to depend on authoritative males, to call one's father or look at his picture constantly, and so on. Intelligence, emotional stability, mathematical ability, and so on, are defined in terms of certain test results that occur when those tests are administered in certain ways. For example, anxiety is defined as a certain range of scores on the Taylor-Johnson test, a person is intelligent if and only if he passes a certain intelligence test, and so on.

Again, our purpose is not to investigate the weaknesses of operationalism.[15] Later in the chapter we will examine how operationalism can be used in integrating science and theology. For now we may note that, like phenomenalism, operationalism is reductionist in spirit. Phenomenalism reduces regular-sized entities like trees and unobservable entities like magnetic fields to nothing more than a set of actual or possible tree experiences (or magnetic-field experiences, e.g., experiences of iron filings). Similarly, operationalism reduces length, intelligence, and the like to nothing more than actual or possible laboratory operations, mathematical operations, and their recording in a notebook. It is one thing to claim, for example, that the color red can be tested for by looking for a certain meter reading of light wavelengths, or even that the color red is somehow *caused* by those light waves.[16] But it is another matter to claim that redness, which appears to be a quality, is identical to a set of wavelengths (a quantity), which in turn can be defined by a set of operations on a laboratory instrument.

In any case, operationalism represents an instrumentalist alternative to a realist understanding of theoretical (and observational) concepts, and while there are very few strict operationalists today,

14. There are different forms of behaviorism. Two brief analyses and critiques of various forms of behaviorism are Joseph Margolis, *Philosophy of Psychology* (Englewood Cliffs, N.J.: Prentice-Hall, 1984), pp. 34–47; Mary Stewart Van Leeuwen, *The Person in Psychology: A Contemporary Appraisal*, ed. Carl F. H. Henry (Grand Rapids: Eerdmans, 1985), pp. 107–40.

15. Cf. Hempel, *Aspects of Scientific Explanation*, pp. 123–33; Harre, *Philosophies of Science*, pp. 76–80.

16. Even this may be mistaken. For a defense of the relative autonomy of colors as basic constituents of the world, see Keith Campbell, *Metaphysics: An Introduction* (Encino, Calif.: Dickenson, 1976), pp. 61–74.

operationalism continues to have a strong influence on scientific practice by expressing itself in the tendency to define a concept in operational terms and then treat it as nothing more than its empirical import expressed in that operational definition.

PRAGMATISM

A third instrumentalist school of thought is pragmatism. As I will use the term, *pragmatism* means, roughly, the idea that science does not give us a progressively truer picture of the world, but aims at giving us theories that work in solving the problems before it. In fact, truth (understood as correspondence with a theory-independent world) is irrelevant for science. Scientific theories allow us to predict phenomena, devise technology, improve experimental technique, represent phenomena in a simple, economical way, and so forth. But theories can (and often do) work without being true or approximately true, and the truth of a theory is basically an irrelevant factor in determining whether it will work.

Two main pragmatists currently writing on the nature of science are Nicholas Rescher[17] and Larry Laudan.[18] In my opinion, Laudan is the main antirealist presently involved in the dispute about science, and he has offered the most persuasive alternative to scientific realism currently available. Thus, I will limit my remarks in this section to an overview of Laudan's position.

According to Laudan, any philosophy of science must take into account three key facts that surface from studying the actual history of science itself. Science is not some disciplinary abstraction or matrix with a specific approach to the world called "the scientific method" (Laudan denies that there is any such thing). Rather, science is a series of events (e.g., discoveries, experiments, and so on) and so should be treated as a creature living through time. Only a philosophy of science adequate to the actual history of science will be able to do the job required of it—provide an accurate understanding of how science has really progressed through the centuries. The three key facts that surface from studying the history of science are:

17. See Nicholas Rescher, *The Limits of Science* (Berkeley: University of California Press, 1984).

18. See Larry Laudan, *Progress and Its Problems* (Berkeley: University of California Press, 1977); *Science and Values: An Essay on the Aims of Science and Their Role in Scientific Debate* (Berkeley: University of California Press, 1984); "A Confutation of Convergent Realism," *Philosophy of Science* 48 (1981): 19–49; "Explaining the Success of Science: Beyond Epistemic Realism and Relativism," in *Science and Reality*, ed. James T. Cushing, C. F. Delaney, and Gary Gutting (Notre Dame: University of Notre Dame Press, 1984), pp. 83–105.

First, the history of science reveals two puzzles requiring explanation—agreement and disagreement. The scientific community often undergoes periods when there is widespread agreement about the adequacy of some scientific paradigm (e.g., Newton's picture of space, time, and motion was agreed on for a long time). The scientific community also frequently undergoes periods when there is major debate about what constitutes an adequate scientific paradigm. In fact, in almost any period in the history of science there will be a number of examples of agreement and disagreement.

Second, science has shown itself to be an objectively rational, progressive discipline, but its progress has had little to do with truth or approximate truth. No one yet has spelled out adequate theories of truth or reference (see chapter 4) to let us know what truth or reference is or how we know when we have the truth or have successfully referred to some theoretical entity. The notion of approximate truth is hopelessly vague, several theories in the history of science have been epistemologically successful (they embodied several epistemic virtues, e.g., simplicity, clarity, predictive success) and later falsified, and several theories realists now take to be approximately true were notoriously unsuccessful vis-á-vis their competitiors for long periods of time. Furthermore, later scientific theories in some domain of study often replace earlier theories instead of refining them. For these and other reasons, truth (or approximate truth) is irrelevant to understanding the progress of science.

Third, any philosophical account of science must give adequate weight to the role of conceptual problems in theory formation and evaluation.

In light of these three features of the history of science, Laudan offers a problem-solving model of scientific rationality and progress wherein progress involves developing theories of increasing problem-solving effectiveness, and the rationality of science is constituted by choosing to use the most effective problem-solving theories. Let us probe Laudan's model of science in more detail.

The nature of scientific problems.   Science seeks to solve two basic kinds of problems: empirical problems and conceptual problems. An empirical problem is anything about the natural world that strikes us as odd and in need of explanation. Examples are the periodic movement of the planets or the fact that the pressure and temperature of a gas are related. There are three kinds of empirical problems: unsolved problems (those not solved by any theory), solved problems (at least one of a set of competing theories has solved it, perhaps several have, even if the solutions are not entirely adequate or do not last as permanent solutions), and anomalous problems (those that a particular theory has not solved but at least one of its rivals has solved; anomalies are not decisive against a theory, but merely raise some doubt against it[19]). A solution to a problem does not need to use an explanation that truly describes real entities or processes in the world. At best, it must only be thought to do so, and even this is not necessary if the scientists working on the problem are not realists. Truth is not relevant to a solved problem. All one needs is a model or explanation that removes the puzzling nature of the problem and makes it a clearer or more expected aspect of the world.

A second type of problem is a conceptual problem that involves the clarity or intelligibility of a scientific theory so far as its structure or conceptual apparatus is concerned. There are two main types of conceptual problems. First, there are internal conceptual problems, which arise when a theory exemplifies circularity, vagueness, internal inconsistency, and so forth. If earlier models of Darwinian evolution used the "survival of the fittest" in a circular way, if the wave/particle theory of light seems to involve a contradiction, then these count as (not necessarily decisive) internal conceptual problems for those theories. Second, there are external conceptual problems. These arise for some scientific theory, can be supported by rational argumentation, and arise in some discipline outside of science (e.g., philosophy, theology, or mathematics). By including external conceptual problems as legitimate aspects of science itself, Laudan broadens scientific methodology and rationality to include, as part of science itself, metaphysical, theological, and philosophical issues. Thus, Laudan holds that any attempt to draw a clean line of demarcation between science and nonscience, especially when that line excludes theological considerations from scientific theory evaluation, will be naive and out of touch with the way science has developed throughout its history.

Problem solving, scientific progress, scientific rationality. A problem is solved when puzzlement regarding it is removed; that is, when a theory allows us to see why a particular phenomenon is not unusual but is normal and to be expected. Theories of celestial motion remove our puzzlement about solar eclipses and make them understandable and predictable.

Rival theories (e.g., Newton versus Einstein, or neo-Darwinianism versus punctuated equilibrium theory versus creationism) can be compared with each other for their relative problem-solving effectiveness.

19. I have used this point in discussing issues involved in the rational appraisal of the biblical doctrine of inerrancy. See J. P. Moreland, "The Rationality of Belief in Inerrancy," *Trinity Journal* 7 (Spring 1986): 75–86.

Such comparisons often are very difficult and may not yield a clear winner in each case but, at least in principle, they can be made. Two rivals can be compared for their relative success in solving problems (e.g., the relative number and importance of the empirical problems each one does or does not solve, the number and significance of anomalous problems—problems one rival solves but the other does not, and the relative number and significance of conceptual problems). Two rivals also can be compared for the relative rates at which they have solved the various problems posed to them.

It should be clear why such comparisons are often difficult. One rival may solve conceptual problems better while the other may solve a greater number of significant empirical problems. How do we devise a scale for weighing which are more important? Further, some phenomenon may be a problem for one rival but not for another. For example, conceptual problems about the aether were unsolved issues for Newtonian theories of motion through absolute space, but they were not problems for Einstein's theory of relativity, which abandoned the existence of aether altogether. Again, motion was not natural in Aristotle's picture of the universe, and examples of motion posed problems in need of explanation. But on Newton's picture of the universe, motion was natural and only changes of motion posed problems in need of solution.

There is a lesson to be learned from the various subtle nuances of theory comparison and evaluation. The picture of scientific theory evaluation in which two theories are judged completely in virtue of some crucial experiment that verifies one and falsifies the other is a naive picture of how theories have been, and should be, evaluated. More specifically, the relative merits of different forms of evolutionary theory versus creationism must take into account the subtlety of theory comparison and evaluation, especially the role of conceptual problems, if the debate over evolution and creation is to be conducted rationally.

Laudan holds that the basic unit of scientific progress and rationality is the solved problem. Scientific progress does not consist in the progressive convergence on a truer and truer picture of the world. Rather, it is a measure of the relative number, rate, and importance of the various problems science solves, where science may be understood as an entire discipline or as some specific set of theories within a given area of science.

Scientific rationality is not to be defined in terms of truth; that is, theory A is more rational than theory B just in case theory A is more likely to be true than theory B. Rather, the rationality of science consists in accepting (acting as if some theory were true, even though it most likely is not) the most progressive theories. It is more rational to accept theory A over rival theory B just in case A is more progressive than B.[20]

Thus, progress is more fundamental for Laudan than is rationality, and problem solving is more fundamental still. That is, he defines scientific rationality in terms of scientific progress, not vice versa. A scientific theory is rational to the extent that it shows progress; it is not progressive because it is rational. And he defines progress in terms of problem solving.

At this point a question naturally arises. According to scientific realism, scientific theories usually solve problems because the entities they postulate in scientific explanations really exist and have, at least approximately, the properties ascribed to them by the successful scientific theories. On Laudan's view, how do scientific theories succeed in solving problems?

Why science progresses in solving problems. In order to understand Laudan's explanation of why science progresses in solving problems, it will be helpful to look first at how, according to Laudan, scientific disputes arise.[21] There are three main levels of scientific dispute. The first and lowest level involves factual disputes. These are competing claims about what exists in the world and refer to both directly observable facts and unobservable theoretical entities. Disputes at this level occur when there is a difference of opinion about the existence and significance of the facts themselves.

The second level of scientific disputes concerns methodological rules. These rules vary from highly general ("formulate testable and simple hypotheses," "avoid ad hoc theories"), to intermediate ("prefer the results of double-blind to single-blind experiments"), to very specific

20. The rationality of acceptance is to be contrasted with the rationality of pursuit of a theory. The former involves the actual problem-solving track record of a theory vis-á-vis its rivals; the latter focuses on the potential promise a fairly recent, undeveloped, or untested theory has for solving problems. Thus, it could be rational to pursue a theory but not rational to accept it, if the theory is relatively new and holds promise. While I think recent creationist theories are rational to accept, even one who disagrees with them could still consider them rational to pursue. Someone could object that creationism has been around for a long time, and thus is not a recent theory. But this objection is wide of the mark. Recent creationist theories do have things in common with creationism in the eighteenth century. But there are also significant differences. Creation scientists need to continue to develop their models, and as they do, they can at least claim that their work involves theories that are rational to pursue, even if (in other scientists' judgment) not rational to accept (though creationists certainly will want to defend the stronger claim that creation science models are rational in both senses).

21. Cf. Laudan, *Science and Values*, pp. 23–66.

("make sure to calibrate instrument x against standard y"). They provide constraints or injunctions about the things we should seek or avoid in our laboratory procedures or in our theories themselves.

The third and highest level of scientific disputes is axiological. Here, disputes take place about the aims, goals, and values that science seeks to embody. Thus, the axiological level includes different understandings of the nature of science. Different aims of science or values include truth, simplicity, predictive success, novel predictions and directions for research, internal clarity and consistency, and so forth.

According to Laudan, the standard explanation of why science successfully solves problems and converges from conflict in some domain to widespread consensus is that it uses a hierarchical model of conflict resolution. If there is a dispute among scientists at the factual level, it can be resolved by appealing to shared rules at level 2. That is, if scientific disputants can agree about rules of procedure, then many factual disputes can be resolved by appealing to common rules. If there is a dispute at level 2, it can be resolved at the axiological level by seeking agreement about the aims, goals, and values of science itself. In light of these shared aims, goals, and values, disputes can be resolved by trying to show what rules of procedure are more likely to secure the agreed on values.

For reasons outside our present concern, Laudan rejects this hierarchical model. One problem he points out is that it leaves unanswered the question of how scientists are to resolve disputes at the axiological level. What if one scientist with theory A holds that science should seek simple theories, another with theory B holds that science should seek theories that make novel predictions, and each theory succeeds in light of the respective values of the two scientists? One is simpler and the other makes more successful, novel predictions.

In place of the hierarchical model of dispute resolution, Laudan offers a type of coherence theory of justification that he calls a reticulated model of scientific rationality, pictured in figure 5.2.

We need not attempt a detailed analysis of Laudan's model here,[22] but, in summary, it illustrates that scientific rationality, which involves resolution of disputes and convergence on scientific consensus, which in turn constitutes progress in problem solving, is not to be understood in realist terms. That is, scientific problem solving does not produce progressively truer and truer descriptions of bedrock facts and data in the external world. Neither are such descriptions secured by altering our methods or our aims, goals, and values until we suc-

---

22. For a critique, see Losee, *Philosophy of Science and Historical Enquiry* (Oxford: Oxford University Press, 1987), pp. 110–13, 130–34.

Figure 5.2
Reticulated Model of Scientific Rationality

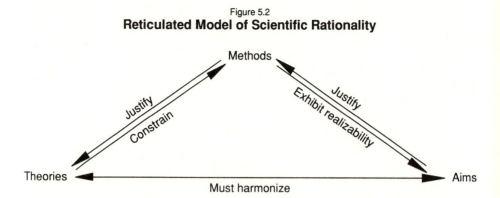

ceed in describing and explaining the facts in an approximately true way. Rather, scientific problem solving involves a complicated interplay among our theories (i.e., our pictures of the observational and theoretical facts), our methodological rules, and our cognitive aims and values.

When there is a dispute at one level, an adjustment can be made at other levels so as to return our entire scientific network of facts, rules, and values to a stable harmony that is internally consistent. For example, if our theory tells us that certain things are facts, then this provides constraints on our methods. If a certain method does not tell us that the facts are what our picture says they should be, we can find other methods that will harmonize with our theories. For example, if one way of calibrating our instruments does not give us the facts we anticipate, we can find other ways of calibrating them that will. Methods, in turn, can help to justify our theories. Similarly, theories must harmonize with our aims and vice versa (e.g., if we are seeking simple theories, not theories with novel predictions, then we will focus on keeping theories that are simple and the value of novel prediction will be less important). Our aims tend to justify our methods, and our methods help us realize our cognitive aims.

Another way to understand Laudan's view here is as follows.[23] The hierarchical model (which Laudan rejects) can be likened to a pyramid, where the earth, representing scientific facts, is the unchanging foundation on which the pyramid (the methods and aims of science) is erected. Pictured in this way, scientific progress in problem solving

23. See Ernest Sosa, "The Raft and the Pyramid: Coherence versus Foundations in the Theory of Knowledge," in *Midwest Studies in Philosophy V: Studies in Epistemology,* ed. Peter French, Theodore Uehling, and Howard Wettstein (Minneapolis: University of Minnesota Press, 1980): 3–25.

can be understood along the lines of scientific realism. The ground remains the same. The pyramid is build on the stable, unshakable foundation of the ground. Similarly, the growth of scientific theories is based on increasing ability to describe and explain accurately the unshakable ground of scientific facts about the way the world is.

Laudan's model, in contrast, can be likened to a raft floating in midair. If one part of the raft needs repair, we can stand on another part and fix it. We need not stand on the ground. Provided that we are always standing on a part of the raft that is not breaking down, we can always repair other parts without stepping outside the raft. Similarly, when a breakdown occurs in the adequacy of our scientific views regarding facts, methods, or cognitive aims, we can always solve problems in that area by holding steady, for the moment, other parts of our conceptual web. If instability is introduced into our scientific framework, we can adjust other parts of our web to restore order, even though we do not step outside the framework itself to do so.

Hence, for Laudan, science progresses in solving its problems not by progressively securing a truer and truer description of the theory-independent world but by constantly readjusting different parts of the web of theories, methods, and cognitive aims. These successive readjustments keep stability in the web of beliefs and thus solve problems without necessarily giving us the truth. Scientific problems get solved because we work on different parts of our conceptual framework until order and consistency are restored. They are not solved by comparing our theories with the theory-independent world in a better and better way until our mirror of the world causes the problems to disappear.

CONSTRUCTIVE EMPIRICISM

The fourth and final type of instrumentalism we will survey is called constructive empiricism, and its chief proponent is Princeton philosopher of science Bas C. van Fraassen.[24] According to van Fraassen, scientific realism is the view that scientific theories aim to give us a literally true story of what the world is like, and acceptance of a scientific theory involves the belief that it is true.

For a number of reasons, van Fraassen rejects a realist notion of science, but one of his reasons is particularly instructive. Scientific realism, he claims, smacks of metaphysics, and inflationary meta-

---

24. Bas C. van Fraassen, *The Scientific Image* (Oxford: Oxford University Press, 1980); "To Save the Phenomena," in *Scientific Realism*, ed. Jarrett Leplin (Berkeley: University of California Press, 1984): pp. 250–59; Paul M. Churchland and Clifford Hooker, eds., *Images of Science: Essays on Realism and Empiricism, with a Reply from Bas C. van Fraassen* (Chicago: University of Chicago Press, 1985).

physics at that. In other words, if one takes a scientific-realist approach to science, then one engages in metaphysical speculation that is virtually identical in argument form (i.e., unseen causes are postulated to explain observable phenomena) to the arguments medieval philosophers like Aquinas advanced to prove the existence of God; involves postulating unobservable essences, properties, dispositions, causal powers, and so forth that are taken to be features of theoretical entities that are causally responsible for observable phenomena. Such postulations go beyond the observational evidence. That evidence does not select one and only one characterization of the theoretical entities. Thus, several different, metaphysical characterizations of those entities are empirically equivalent regarding the observational data. Hence arguments about the existence and nature of those theoretical entities involve metaphysical speculation.

An illustration may be helpful. Prior to the advent of modern chemistry, the chemical interaction of various substances like gold was explained by postulating various metaphysical notions of those substances. Thus, gold was seen as a substance with a substantial form or essence defining the very nature of goldness and was thought to have various causal powers or dispositions (i.e., lawlike tendencies to act in certain ways if certain circumstances obtained; gold has the capacity to dissolve if it is immersed in *aqua regia*), and these properties and dispositions of goldness were occult (unobservable) aspects of gold postulated to explain its behavior.

Chemists like Robert Boyle (1627–1691) eschewed such metaphysical speculation and argued that chemical change was due to purely mechanical interactions of force, collision, and rearrangement of tiny, corporeal particles called atoms by virtue of their motion, size, figure, and so on. According to van Fraassen, Boyle was right to try to avoid postulating real properties, essences, and dispositions of chemical substances. But these problematic, metaphysical entities surface all over again in the atoms themselves. How are we to explain the uniform behavior of the class of carbon atoms? The realist will do so by saying that carbon atoms have a certain essence or set of real properties and certain dispositions to behave in certain ways given certain circumstances (e.g., part of the nature of a carbon atom is that it has a disposition to fill its outer electron shell).

So a realist understanding of science commits one to at least the following: 1) the postulation of metaphysical entities, which makes metaphysics a part of empirical science; 2) rational, metaphysical speculation that goes beyond and is underdetermined by the empirical data (i.e., different metaphysical views will be equally compatible with the data); 3) a form of argumentation that uses a leap of intellect beyond observational data to postulate causal entities and causal connections that allegedly explain those data, a form of argumentation virtually identical to that used in natural theology.

All of these are unacceptable to van Fraassen. Science is indifferent about whether electrons really exist and cause things, and the proper scientific stance toward unobservable theoretical entities is agnosticism. The proper scientific attitude is nominalism (i.e., the denial of the existence of real properties, essences, and dispositions) and empiricism.

In place of scientific realism, van Fraassen proposes what he takes to be a more adequate philosophy of science—constructive empiricism. Van Fraassen calls his view "constructive" because he sees the aim of scientific activity to be the construction of theoretical models, not the discovery of truth concerning unobservables. It is called "empiricism" because the main goal of science is to "save the phenomena," that is, to develop empirically adequate theories that accurately describe and predict observational phenomena. Let us look at constructive empiricism in more detail.[25]

According to van Fraassen, there is a difference between accepting a theory and believing it. To believe a theory is to adopt a scientific realist attitude and believe it is true, that is, that some model or picture accurately represents the world of unobservable, theoretical entities. In contrast, accepting a theory involves at least two things.

First, acceptance involves belief that the theory is empirically adequate, that is, that it is true regarding what it says about what is observable by us. A model or picture of the theory will be isomorphic with the observational world. Put differently, if I accept a theory, that means that I believe that those aspects of it that describe or predict what can be observed do so in an accurate, true way. What the theory says we should observe is what we observe. A theory draws a picture of the world and certain aspects of that picture refer to what is observable. If one accepts the theory, one asserts that the picture is accurate in those observational aspects. So accepting a theory involves no more belief than that the theory accurately represents observational phenomena. The realist, in contrast, believes a true theory to be a true description of the observable *and unobservable* world. Con-

---

25. There are two main aspects to van Fraassen's philosophy of science: the realist debate about the relationship of a scientific theory to the world, and the nature of scientific explanation. We will not discuss the latter. For a discussion and critique of van Fraassen's theory of explanation, see Karel Lambert and Gordon G. Brittan, Jr., eds., *An Introduction to the Philosophy of Science*, 3d ed. (Atascadero, Calif.: Ridgeview, 1987), pp. 34–42, and a review of *The Scientific Explanation* by Martin Gerwin in *Canadian Journal of Philosophy* 15 (June 1985): 363–78.

structive empiricism takes the weaker epistemic attitude and only accepts the theory (i.e., believes that it is empirically adequate). Thus, the constructive empiricist shows the restraint necessary to deliver science from excess metaphysical baggage involved in postulating unobservable, theoretical entities.

Second, accepting a theory involves agreeing to act *as if* the theoretical entities postulated by it are real. One adopts that stance as a pragmatic posture that facilitates the scientific search for new empirically observable correlations in the world. One commits oneself to the research program implicit in the theory, to continuing the dialogue with nature contained in the conceptual scheme of the theory accepted. Thus, we may notice certain empirically observable regularities about gold. We can postulate gold atoms with such-and-such structures if such a postulation helps to guide us in the discovery of new empirical regularities involving gold. The theoretical entity, gold atoms in this case, is a useful, imaginative picture whose sole value is that it may be fruitful in guiding us to the discovery of new, observable correlations and the abandonment of old correlations. But the theoretical entity need not really exist at all.

Thus, those theoretical aspects of a theory that are unobservable are pragmatically justifiable aspects of the theory that may or may not help to guide us in our search for new observational correlations. We act as if the theoretical entities exist, but we do not really believe they exist. We accept those entities (i.e., we use them pragmatically to achieve greater empirical adequacy). The basic value of the unobservable aspects of a theory is that they facilitate the development of more accurate empirical descriptions and predictions of observational phenomena.

At this point, a scientific realist will object, "Why does the pragmatic acceptance of some theoretical entity lead to greater empirical accuracy regarding our observational descriptions and predictions?" Realists have an answer to this question, namely, that the theory refers to real entities that are causally responsible for the observational data and describes these entities in an approximately true way. Therefore we can expect research guided by the postulation of these entities to lead to new observational data. But no such explanation is available to the constructive empiricist, and the empirical success of a theory must remain a miracle or happy coincidence.

Van Fraassen responds in three ways. First, the sole value of the theoretical, unobservable postulates of a theory is the empirical observations they lead to. Theories are always underdetermined by observational data; there is always a potentially infinite number of theories equally compatible with observational data, and the dispute about which theory is the true mirror of the unseen world is an un-necessary metaphysical dispute outside the bounds of science. The scientist is agnostic about such questions and is interested basically in the observational consequences of the theory. For example, Newton distinguished real from apparent motion. The former, he said, is the motion of a body with respect to the theoretical entity absolute space (and time). The latter is the motion of a body relative to some other body, perhaps the sun, the fixed stars, or, more importantly, observers on earth. Newton also believed that the center of gravity of the solar system is at rest in absolute space. However, the observational ap-pearances of motion would be unaffected if the center of gravity of the solar system moved through absolute space at any constant velocity. Hence a potentially infinite number of conflicting theories arise that differ regarding the theoretical behavior of the center of gravity of the solar system (it could be moving at any velocity v, including $v = 0$) but that are all empirically adequate. In fact, one could deny the ex-istence of absolute space and time and accept Newton's theory anyway by holding that it is empirically adequate but not true in its theoretical aspects. Which stance one adopts is not relevant. What is relevant is the empirical adequacy of each model, and if they all agree in this regard, a choice between them is of little scientific interest.

Second, van Fraassen argues that the demand for explanation must stop somewhere. If we explain the regularities of observable, chemical phenomena in terms of the behavior of unobservable atoms, then we will need to postulate a new set of regularities governing the behavior of these new entities. Some set of regularities will always have to be taken as brute, unexplainable givens if we are to avoid a vicious infinite regress of explanations. So even the realist must embrace the existence of basic regularities that have no explanation, and the constructive empiricist would rather accept observable regularities as brute facts than hidden, unobservable regularities.

Finally, according to van Fraassen, science is a type of biological phenomenon (an activity of one kind of organism that facilitates its interaction with the environment), and a Darwinian model of theories themselves may be the best view of them. Two contrasting accounts can be given for why a mouse runs from its enemy, the cat.[26] According to Saint Augustine, the mouse perceives that the cat is an enemy. What is postulated here is the adequate mirroring of the world by the

26. See van Fraassen, *Scientific Image*, pp. 39–40. For a criticism of van Fraassen, see Richard Boyd, "Lex Orandi est Lex Credendi," in *Images of Science*, ed. Churchland and Hooker, pp. 23–28. Van Fraassen has responded to Boyd on pages 282–84 of the same volume.

mouse—it correctly grasps the relation of enmity and survives. According to Darwinism, the mouse does not survive because it correctly grasps the enmity relation it sustains with the cat. Rather, mice that don't run die; only mice that run survive.

Van Fraassen applies this to scientific theories as follows:

> In just the same way, I claim that the success of current scientific theories is no miracle. It is not even surprising to the scientific (Darwinist) mind. For any scientific theory is born into a life of fierce competition, a jungle red in tooth and claw. Only the successful theories survive—the ones which *in fact* latched on to actual regularities in nature.[27]

SUMMARY

This completes our survey of the various rational nonrealist or instrumentalist positions in the philosophy of science. Phenomenalism, operationalism, pragmatism, and constructive empiricism differ in various ways, but they agree on two important points: first, that science is an objectively rational set of disciplines, rationality is a normative, objective notion, and conceptual relativism is false; second, that a realist understanding of science is false.

It is time now to turn to an investigation of a more radical stance about science—nonrational nonrealism.

Nonrational Nonrealism

Nonrational nonrealists agree with instrumentalists in denying scientific realism. Science does not progress toward ever more accurate pictures of the theory-independent world. However, nonrational nonrealists add to their rejection of scientific realism their view that rationality is not an objective, normative notion. They are, in short, conceptual relativists.

A number of thinkers can be classified as nonrational nonrealists: W. V. O. Quine,[28] Paul Feyerabend,[29] Hilary Putnam,[30] Richard Rorty,[31]

27. Van Fraassen, *Scientific Image*, p. 40.

28. See W. V. O. Quine, *Ontological Relativity and Other Essays* (New York: Columbia University Press, 1969); *From a Logical Point of View*, 2d ed., rev. (Cambridge: Harvard University Press, 1980). For a helpful overview of Quine's thought, see Christopher Hookway, *Quine* (Stanford, Calif.: Stanford University Press, 1988).

29. Paul Feyerabend, *Against Method* (London: New Left Books, 1975). A good overview of Feyerabend's thought is W. H. Newton-Smith, *The Rationality of Science* (Boston: Routledge and Kegan Paul, 1981), pp. 125–47.

30. Hilary Putnam, *Philosophical Papers, Volume 3: Reason, Truth, and History* (Cambridge: Cambridge University Press, 1981); "Why There Isn't a Ready-Made World," *Synthese* 51 (1982): 141–67; "Realism and Reason," *Proceedings and Addresses of the American Philosophical Association*, 50 (6): 483–98.

31. Richard Rorty, *Philosophy and the Mirror of Nature* (Princeton, N.J.: Princeton University Press, 1979).

and Thomas Kuhn.[32] In this section we will take Kuhn as our representative of nonrational nonrealism and focus on his philosophy of science. This is not to imply that all members of this school agree with Kuhn, but he is fairly representative and an overview of his views will enable us to understand the main features of nonrational nonrealism.

KUHN AND THE HISTORY OF SCIENCE

Kuhn was one of the first philosophers of science to emphasize that a philosophy of science must fit the actual history of science and the practice of scientists. On Kuhn's view, if one takes the actual history of science into account, then three important concepts surface as an aid to understanding that history: paradigms, normal science, and crisis-resolving scientific revolutions.

Paradigms and the epistemic authority of science. The epistemic authority of science does not reside in the fact that there is a thing called the scientific method (Kuhn denies that there is such a thing) practiced by all and only scientists. Rather, the authority of science resides in the scientific community seen as a group of practitioners who share a paradigm. This paradigm tells the community what are good and bad science, significant and irrelevant problems, and criteria for acceptable solutions to problems.

What exactly is a paradigm? Kuhn was unclear about this, and his writings contain as many as twenty-two different uses of the term. Nevertheless, his understanding of a paradigm becomes fairly clear in light of his uses of the concept. The following are some of his major uses of it:

a. paradigm as *Weltanschauung*
b. paradigm as theory
c. paradigm as pedagogical tool and educational goal
d. paradigm as gestalt
e. paradigm as constitutive of nature
f. paradigm as what the members of a scientific community share
g. paradigm as disciplinary matrix
h. paradigm as shared exemplar

32. Kuhn, *Structure of Scientific Revolutions*, 2d ed., enlarged (Chicago: University of Chicago Press, 1970); *The Essential Tension* (Chicago: University of Chicago Press, 1977); "Logic of Discovery or Psychology of Research?" and "Reflections on My Critics," in *Criticism and the Growth of Knowledge*, ed. Imre Lakatos and Alan Musgrave (Cambridge: Cambridge University Press, 1970), pp. 1–23 and 231–78, respectively. For overviews of Kuhn with extensive bibliography, see J. P. Moreland, "Kuhn's Epistemology: A Paradigm Afloat," *Bulletin of the Evangelical Philosophical Society* 4 (1981): 33–60; Gary Gutting, ed., *Paradigms and Revolutions* (Notre Dame: University of Notre Dame Press, 1980).

Definition a sees a paradigm as an entire world view. Physicalism as a world view would be an example. Definition b emphasizes the paradigm as a theory of the world such as the Copernican theory of the solar system or Lavoisier's oxygen theory, which opposed phlogiston theory. Definition c emphasizes the role of a paradigm in teaching science; namely, it introduces students to the proper ways of viewing subject matter and solving problems. The ideal gas equation, $PV = nRT$, is a standard way of teaching chemistry students to analyze gases, even though, technically, it only applies to ideal situations and not to the real world (the Van Der Waals equation is a more accurate refinement of the equation).

Definition d points out that a paradigm is a way of picturing or seeing the world; that is, it provides an interpretive perspective for organizing sensory experiences. We see red litmus paper as an indicator of an acid, which in turn is characterized in a certain way by the paradigm. Definition e implies that a paradigm tells what is and is not real. To be is to be an observational or theoretical entity in the paradigm. The world is what the paradigm says it is—nothing more and nothing less.

Definition f brings out the sociological dimension of a paradigm. A paradigm unites the scientific community into one group of practitioners. Thus, two different communities with two different paradigms, Ptolemy versus Copernicus or Lavoisier versus Priestly, live in different worlds and practice different sciences.

Definition g implies that a paradigm defines a discipline (to be a molecular biologist is to have a certain shared theory of molecular biology) by specifying the common possession of its practitioners (this is similar to definition f). It also refers to the role of a paradigm in ordering the phenomena of science (e.g., the periodic table orders the various observational chemical phenomena), specifying metaphysical beliefs of science (e.g., heat is the kinetic energy of the constituent parts of bodies), and ordering epistemic values (e.g., prediction is to be preferred to simplicity). Finally, definition h brings out the role of a paradigm (especially in standard science textbooks) in specifying good ways to view and solve problems so that all members of the community see things the same way (e.g., inheritance phenomena are to be understood by using Mendel's law).

In sum, a paradigm is a universally recognized scientific achievement that, for a time, provides model problems and solutions for a community of practitioners. It is a way of looking at the world. In fact, it constitutes the world, and it provides rules of the game, creates a community, defines good science, and provides the ultimate locus of science's rational authority.

Normal science. The history of science exhibits periods when science can be conceived of as normal. Normal science is characterized by the existence of a universally accepted paradigm. For example, the period when Newton's theories were universally accepted (prior to Einstein) would count as a period of normal science. In such a period, the paradigm is immune from criticism. Scientists spend their time extending the range of the paradigm's application, making the paradigm more precise and accurate (which often involves devising new instruments), and clarifying its conceptual apparatus. All of this work is done from within the paradigm itself. Anomalies (problems that the paradigm does not clearly solve) are dealt with in an ad hoc manner by readjusting the paradigm, denying their significance, disputing their existence (e.g., by arguing about the accuracy of the anomalous data), and so on. During periods of normal science, textbooks tend to portray the paradigm to new students as unproblematic.

Crisis and scientific revolution. The history of science also exhibits a number of periods of what Kuhn calls crisis. These are characterized by the presence of various rival paradigms, no single paradigm being universally accepted. They are caused by the existence of a number of troublesome anomalies inadequately solved by the main paradigm constituting the period of normal science preceding the time of crisis.

Rival paradigms often are introduced by small "rebel" bands of scientists. Kuhn says these groups often are formed by scientists new to the field, since they have not worn the spectacles of the accepted paradigm long enough to be kept from seeing the world differently. These groups of scientists often are ostracized and criticized by the main body of scientists for not being good members of the community—they do not practice good science, perhaps they are not even practicing science at all, and they are engaging in unjustifiable speculation. Yet the existence of these groups is crucial for the progressive changes of science through its history.

It is interesting to view the present resentment of the scientific community toward creation science in this light. I am not claiming that everything in the creation/evolution debate can be reduced to the scientific tendency to ostracize rebel groups that challenge the status quo. But Kuhn has clearly shown that such behavior is frequent in the history of science and that it can be counterproductive, since rebel groups stimulate science to develop new paradigms. Thus, at least part of the creation/evolution debate can be seen in this sociological light, and this helps to explain the emotional baggage and ad hominem texture of so much of the debate. It also serves as a warning to anyone who naively dismisses creationism by claiming that it is unhelpful to science. It may be helpful (by identifying real anomalies) even if it is false. In this case, the common slogan is more than a cliche: the one who fails to learn from history is destined to repeat it.

Returning to Kuhn, a period of scientific crisis causes a scientific revolution. A scientific revolution occurs when a new paradigm defeats its rivals, becomes the new, universally accepted view and ushers in a new period of normal science. In such revolutions the old paradigm is not gradually and cumulatively refined into the new paradigm. Rather, a paradigm shift occurs. The new paradigm completely replaces the old one, even if it retains some of the old terms, in which case, the terms are used equivocally. According to Kuhn, Einstein did not refine Newton, his theory replaced Newtonianism, even though Einstein still used terms like mass, space, and so forth.

Thus, the history of science is one of periods of normal science followed by crisis, which gives way to a revolution in which a paradigm shift occurs and ushers in a new period of normal science. The history of science, therefore, is not what the realist claims it to be—a history of new theories (usually) refining old ones, preserving them as limiting cases, and hence advancing cumulatively toward truer and truer pictures of the world. Rather, it is a history of jerky replacements. Old theories are abandoned, new ones are embraced. The same will happen to our current theories. Thus, the history of science warns us against believing that science, present theories included, is a rational, truth-obtaining enterprise.

KUHN AND THE EPISTEMOLOGY OF SCIENCE

For Kuhn, not only the history of science but also the epistemology of science refutes scientific realism. Two aspects of Kuhn's thought are central to his understanding of the epistemology of science.

Facts, neutral data, and the given.   According to Kuhn, there are no neutral, uninterpreted facts. Observations themselves are paradigm-dependent. Different people with different paradigms see different worlds. Ptolemy saw a moving sun, Copernicus saw a stationary sun. Seeing involves seeing *as* or seeing *that*, and this shows that there is an irreducible dependence of observation upon theory.

Another way to put this is in terms of "the given." Foundationalists in epistemology have identified the given as what is directly present to the conscious, knowing subject and serves as an unchanging foundation for justifying all other knowledge claims.[33] If I am gazing around

33. For those unfamiliar with the epistemological debate about foundationalism, two introductory discussions of this issue are J. P. Moreland, "Foundationalism and Coherentism: A Comparison," *Bulletin of the Evangelical Philosophical Society* 9 (1986): 21–30; Robert Audi, *Belief, Justification, and Knowledge* (Belmont, Calif: Wadsworth, 1988), pp. 80–100.

the room, the given would be the table and the apple directly present to me, or perhaps a circular-shaped, brown color patch with a circular-shaped red color patch lying immediately above it. According to foundationalism, two different theories can be adjudicated because each must interpret the same observational data, namely, the given.

Kuhn denies the existence of such a given. He postulates the existence of incoming sensory stimuli impinging on our sense organs (thus, he avoids solipsism in a Kantian manner), but before such stimuli become observational data, they must first be organized into perceptions by the knowing subject, who sumsumes the stimuli under his theoretical concepts. There is no 1:1 correspondence between stimuli and sense data.

Furthermore, there is no neutral, theory-independent observational language for expressing observations. Such a language would enable two rival paradigms to express the observational data in a common language and would serve, at least in principle, as a common court of appeal in comparing the relative merits of the two paradigms vis-á-vis the observational facts expressed in that observational language. But observational language is itself paradigm-dependent, and there is no way to translate the observational language of one paradigm into that of another. "The pointer is at 30 miles per hour" may be uttered by a Newtonian and an Einsteinian, but they equivocate, that is, they mean different things by *mile, hour,* and *travel through space.*

The upshot of Kuhn's views is that truth, understood as correspondence with a theory-independent world, is irrelevant for science and inaccessible. We simply cannot get outside our theories to see if they correspond to the world. In fact, the notion of a fact, or of the world, is itself paradigm-dependent. There are only facts-in-theory, there is only a world-of-theory. There is no such thing as a fact or a world *simpliciter.* Practitioners of different paradigms live in different worlds.

Incommensurability, theory choice, and gestalt switches. If two paradigms can be rationally compared to each other, at least in principle, such that one is judged more rational than the other, then they are commensurable.

Kuhn holds that rival paradigms are incommensurable. They cannot even be compared with each other for rational assessment. This is because nothing outside the paradigms can serve as common ground for such an assessment. Two rival theories equivocate about the meanings of the terms they employ. Thus, they are literally talking about different things. Facts and criteria for theory choice (e.g., simplicity, scope of explanation, accuracy, fruitfulness, consistency) are all paradigm-dependent with regard to both meaning and relative importance. One paradigm will mean one thing by *simplicity,* another will mean something else.

For example, aether theory about space will seek the most simple theory of what the aether is like; relativity theory will seek simplicity in abandoning the aether altogether. So each theory means something different by simplicity (e.g., the smallest number of properties of an entity versus the smallest number of basic entities). Even if two rival paradigms meant the same thing by data, simplicity, and the like, different pradigms would rank the relative importance of these epistemic values differently. Either way, there is no neutral ground for comparing paradigms. Paradigms are the whole show; everything else depends on them.

This means that the rationality of theory choice is a relative notion. What is rational for one theory (e.g., a simple model of the aether) is not rational for another; what is a problem-posing question for one theory (e.g., the null results of the Michelson/Morley attempt to measure motion through the aether) is not a problem for a rival theory (the null results were a problem for aether theory but not for relativity theory). In fact, in scientific theory assessment and rationality, the rational/nonrational or objective/subjective distinction breaks down. There is no way to draw a line between these two, and rationality is itself a relative notion.

When a scientist or a community of scientists changes from one paradigm to another, the change is not the product of a cumulative addition of more and more observational data to the old paradigm, gradually transforming it into the new paradigm. Instead it is a gestalt switch. (A gestalt is a way of seeing, an observational grid, a perceptual approach or perspective.) Thus, a paradigm shift involves replacing one way of seeing the world with another.

A gestalt switch is usually illustrated by the (infamous) duck-rabbit in figure 5.3. If you view the diagram from one perspective or perceptual gestalt, the dot can be seen as the eye of a duck. If you shift your perspective, the dot can be seen as the eye of a rabbit. The pattern of

Figure 5.3
The Duck-Rabbit

the sensory data is organized differently depending on the perceptual gestalt one takes in approaching the diagram. Notice that the shift from one perspective to the other does not occur gradually by the accumulation of more and more perceptual information. It occurs all at once. Now I see it as a duck, now as a rabbit. What counts as an observation—the observation of a duck eye or of a rabbit eye—is determined by the gestalt.

In the same way, says Kuhn, theory choices in science occur all at once. In a scientific revolution, an old gestalt is abandoned and a new one is adopted. Such a shift occurs not piecemeal but all at once. It involves a conversion, a change from one perspective to another.

In sum, scientific rationality is a relative notion. What is rational for one paradigm is not necessarily so for another. In fact the rational/nonrational distinction is itself arbitrary. Conceptual relativism is an appropriate epistemological stance in keeping with scientific activity. Furthermore, scientific realism does not do justice to the epistemology or the history of science, the latter being a history of successive replacements of entire gestalts, not a history of cumulative refinement and preservation of earlier theories by later ones in progressive growth toward the truth about a theory-independent world.

I have criticized Kuhn elsewhere and it is not relevant to our purposes to sustain a detailed criticism of his views here.[34] Suffice it to say that if Kuhn is given a strong reading, one that emphasizes his conceptual relativism, the complete denial of objective facts, and the all-embracing nature of a paradigm, then his views must be judged inadequate, for his position is self-refuting in several ways. In this context, it is interesting to see the difficulties Kuhn had in "arguing" against Karl Popper's views in the philosophy of science.[35] Popper's paradigm in the philosophy of science was itself a rival to Kuhn's, and when faced with the debate, Kuhn offered this query:

How am I to persuade Sir Karl, who knows everything I know about scientific development and who has somewhere or other said it, that what he calls a duck can be seen as a rabbit? How am I to show him what it would be like to wear my spectacles when he has already learned to look at everything I can point to through his own?[36]

If Kuhn is to be consistent, one way not to do it would be to maintain that he, Kuhn, had succeeded in explaining the actual facts of the

34. Moreland, "Kuhn's Epistemology."
35. This argument is discussed in Kuhn, "Logic of Discovery or Psychology of Research?"
36. Ibid., p. 3.

history of science better than had Popper, since according to Kuhn's theory there are no actual facts of the history of science.

But Kuhn can be given a softer reading, one that does not push the relativist aspects of his thought. According to this reading, Kuhn has taught us how difficult it is to get at the facts and how central the role of theory is in such an endeavor; how important social perspective and pressure have been in the history of science; and how much of the history of science can be interpreted as a sequence of theory replacements, not refinements.

If the softer understanding of Kuhn is allowed, then his philosophy of science warns us against embracing a naive, overconfident scientific realism wherein observational facts and crucial experiments are seen as the major events in theory testing and where the complexities of theory assessment and theory change in the history of science are not adequately taken into account.

Most realist philosophers of science do take the history of science into account. But I suspect the average layperson or scientist himself embraces scientific realism without knowing the issues involved in such a position or the importance of the history of science itself. When these are considered, several things can be learned. One is that metaphysical and theological concepts always have been a part of science—and a useful part at that. The application of this observation to the current debates about the interplay between science and theology should caution us against any view that postulates that these two domains are entirely different, noninteracting approaches to the world.[37]

An Eclectic Approach to Science

In this and the last chapter we unearthed several objections to rational realism and, in the process, some reasons for accepting it (e.g., it is offered as the best explanation of the progressive success of science). We have also looked at some major alternatives to it. What are we to make of this debate? Should we adopt rational realism or some form of antirealism?

Christians have long been divided on this question. In our day, for example, Christians like Gordon Clark and John Byl adopt instrumentalism as an appropriate philosophy of science for developing a co-

37. For a treatment of different models of integrating science and theology, see J. P. Moreland, *Scaling the Secular City: A Defense of Christianity* (Grand Rapids: Baker, 1987), pp. 200–208.

herent Christian world view.[38] Clark argues that science, with its heavy reliance on sense experience and observation, is unable to penetrate beyond the phenomena and give us knowledge of unseen causes. Thus, he adopts an operationalist interpretation, not of epistemology as a whole (he allows for philosophical argumentation about the nature of ultimate reality) but of science. Science gives us not knowledge or truth but useful generalizations.

Byl's version of instrumentalism puts forth the notion that scientific theories are useful fictions that aid in summarizing, manipulating, and predicting observations and are to be judged by their utility, not their veracity. Thus, when a scientific theory seems to conflict with a scriptural teaching, conflict resolution involves seeing the scientific theory as a useful instrument that, since it is not literally true, cannot conflict with the scriptural teaching.

On the realist side are Christian thinkers like Del Ratzsch and Stanley Jaki.[39] Ratzsch argues that Christianity lends support to realism because, on a Christian view, God created us with sensory and rational faculties appropriate for a true knowledge of creation, including the unobservable aspects of creation. Jaki agrees and adds that science and natural theology proceed in a similar way—both use a bold leap of the intellect beyond sensory phenomena to the postulation of unseen causes responsible for those phenomena. Thus both natural theology and a realist understanding of science reject crude empiricism and use similar structures in arguments to the best explanation.

In my opinion, we are not forced to choose one option or the other. I see no reason why one cannot adopt an eclectic approach to science that adopts a realist/antirealist view on a case-by-case basis. I must confess that I have a good bit more thinking to do on these questions, and it is with some measure of fear and trembling that I offer an eclectic approach with some suggested guidelines for when to adopt realism or antirealism. My main purpose is to help believers interested in the interaction of science and theology to use philosophical considerations in general and, more specifically, realist/antirealist alternatives, in their attempt to develop an overall framework for integrating science and theology into a satisfying Christian world view.

My reasons for eclecticism are varied. I share with Ratzsch and Jaki the conviction that our senses and intellect are in principle adequate

38. See Gordon H. Clark, *The Philosophy of Science and Belief in God* (Nutley, N.J.: Craig, 1964); John Byl, "Instrumentalism: A Third Option," *Journal of the American Scientific Affiliation* 37 (March 1985).

39. Del Ratzsch, *Philosophy of Science* (Downers Grove: Inter-Varsity, 1986), pp. 125–26; Stanley Jaki, *The Road of Science and the Ways to God* (Chicago: University of Chicago Press, 1978).

to infer causal connections and unseen entities behind the phenomena. It also seems to me that some form of correspondence theory is the most rational view of truth and that correspondence is a main aspect of the theory of truth taught in the Bible itself.[40] Further, we all know that we can successfully refer to things in the world, even though we may not have an adequate theory of how we do it—that is, of what reference is and how it works. Skepticism regarding a particular theory of reference should not be taken to imply that we really cannot refer at all. Rather, skepticism about reference is a heuristic device, a foil that urges us to press on in our efforts to understand reference (and meaning).

Again, the observation/theory distinction, though difficult to specify precisely, is still correct and is best viewed as a continuum—clear cases of observation and theory can be identified at each end, and there are several borderline cases in the middle where the line cannot be drawn.

On the other hand, the history of science teaches us several things that favor antirealism: successful theories often turn out false; theories construed as true on a realist view of current science were unsuccessful for long periods of time; theories are often, perhaps always, empirically equivalent in such a way that the observational data do not unambiguously point to one superior theory; nonrational sociological factors influence scientific decision making, and so on. Furthermore, the history of science shows a lot of theory replacement instead of theory refinement, an observation underscored by the difficulty in making precise the observation/theory distinction, how reference is accomplished, and what it means for a later theory to refine and preserve an earlier one in the same domain. These difficulties have allowed us to see more refinement in the history of science than is warranted. And if the history of science contains a lot of theory replacement, a pessimistic induction warrants skepticism about the approximate truth of our currently held scientific theories even though they are successful in many ways.

This is a far too brief rationale for why I think eclecticism is justified. I do think, however, that enough has been said to suggest the kinds of considerations presented in the last two chapters that warrant such a position. And I am not alone in advancing eclecticism as an adequate philosophy of science. Philosopher of science Janet Kourany cites, approvingly in my view, a growing number of thinkers like Michael Gardner who adopt the following:

40. Cf. Norman Geisler, "The Concept of Truth in the Inerrancy Debate," *Bibliotheca Sacra* (October-December 1980): 327–39.

But a question such as this [the realist/antirealist question], about the proper interpretation of scientific theories, should be settled through an examination of particular actual scientific theories rather than in general terms, given the possibility that the question might have different answers for different theories.[41]

In sum, an eclectic approach to the realist/antirealist debate deserves serious consideration and could provide a conceptual framework for developing one's view about the integration of science and theology. For example, when science and a theological statement or biblical interpretation come into conflict, part of the solution may lie in adopting an antirealist view of the scientific statement. This is not the only strategy relevant to conflict resolution, but it is a factor that should be considered.[42]

But how do we know when to adopt an antirealist interpretation of a scientific theory and when to adopt a realist one? I'm afraid that I have nothing that even approximates a full-blown set of criteria for answering this question. But as Kourany suggests, an adequate formulation of eclecticism may need to start not with criteria but with an examination of specific cases of scientific theories.[43] This point should provide added motivation for Christian scientists, philosophers, historians, and theologians to work together in teams on the task of developing a Christian world view of science and Christianity that is sensitive to the scientific, historical, philosophical, and theological aspects of such an endeavor.

With this caveat in mind, here are two criteria for when to adopt an antirealist view of a scientific theory.[44] These criteria should be understood in light of the following. First, the burden of proof is on the antirealist, that is, a scientific theory should be understood along

41. Janet A. Kourany, *Scientific Knowledge: Basic Issues in the Philosophy of Science* (Belmont, Calif.: Wadsworth, 1987), p. 339.

42. It may be objected that this approach begs the question by assuming the truth of the biblical interpretation or theological statement from the outset. One's response to this objection will turn, in part, on one's approach to apologetics. If one is a fideist or a presuppositionalist (roughly, the view that rational argumentation and evidence cannot be offered as epistemic support for Christian theism from some neutral starting point), then one may say that begging the question is not a problem here. If one is an evidentialist, as I am, then one can suspend judgment about the theological or biblical component of the apparent conflict by viewing it as a rationally justified conceptual problem for the scientific theory in question.

43. Cf. Roderick Chisholm, *The Problem of the Criterion* (Milwaukee: Marquette University Press, 1973).

44. For a list of other criteria, see Michael Gardner, "Realism and Instrumentalism in Pre-Newtonian Astronomy," reprinted in *Scientific Knowledge*, ed. Kourany, pp. 369–87.

realist lines in the absence of sufficient evidence to the contrary. Second, by *antirealism* here I mean no more than this: the apparent success of the theory in question should not be taken as evidence of the truth or referential import of the theoretical aspects of the theory. I leave aside the question of what version of antirealism one should adopt in a given case.

An antirealist approach should be taken toward some scientific theory in those cases where the phenomena described by that theory lie outside the appropriate domain of science, or the scientific aspect of some phenomenon is inappropriately taken to be the whole phenomenon itself.

The main insight expressed in this criterion should be clear. It attempts to refute a tendency in scientism toward reductionism and physicalism. When a realist construal of a successful scientific theory is used to advocate reductionism or physicalism, then if the latter are to be avoided, the success of the scientific theory in question should be treated in antirealist terms.

An example would be areas of science that impinge upon mind, life, or various mental states like thinking, being depressed, seeing a red object, or feeling pain in one's tooth. Consider the mind. A recent article in *U. S. News & World Report* gave a popular summary of what many, perhaps most, scientists and philosophers of mind claim.[45] Included in the article were claims that memories are made of nerve cells (i.e., nerve cells are constituent parts that make up memory-stuff); Descartes viewed the mind [sic] as a machine, in which nerves were likened to the plumbing of fountains; and fear, happiness and love are all part of the mind's machinery.

In another work, philosopher of science Paul Churchland claims that a type of mental state, namely, human joy, is really identical to a type of physical state, namely, resonances in the lateral hypothalamus.[46]

---

45. William F. Allman, "How the Brain Really Works Its Wonders," *U.S. News & World Report*, June 27, 1988, pp. 48–54.

46. Paul M. Churchland, *Matter and Consciousness* (Cambridge, Mass.: MIT Press, 1984), pp. 41–42. Later, Churchland claims: "If machines do come to simulate all of our internal cognitive activities, to the last computational detail, to deny them the status of genuine persons would be nothing but a new form of racism." See p. 120. This remark reminds me of an observation made by Howard Robinson about the proper order between philosophy and science, the former being conceptually and epistemically prior to the latter. When this order is reversed, says Robinson, in the interests of furthering a materialist program, highly counterintuitive results follow. See Howard Robinson, *Matter and Sense: A Critique of Contemporary Materialism* (Cambridge: Cambridge University Press, 1982), p. 109. This is one of the reasons why Christians engaging in the science/theology dialogue cannot afford to overlook the insights of philosophy.

Similar observations could be made about the tendency in sociology or psychology to identify various mental states like intelligence, depression, or anxiety with falling within a certain range on some standard test or with certain overt behavior or tendencies to behave (e.g., the behaviorial practice of defining pain as the tendency to grimace and shout "ouch!" when stuck with a pin; pain may be caused by a pin stick and I may, in turn, shout "ouch!" but pain is not identical to these behaviors).

In all of these cases, mental phenomena are being defined so as to identify them with physical phenomena. But mental phenomena would seem to be entities and processes outside the domain of science in contrast to the physical entities and processes that stand in various causal relations with those mental phenomena.

I have defended the irreducibility of the mind to matter elsewhere, and it is not important to do so here.[47] Suffice it to say that if some form of mind/body dualism is true, then when scientists devise successful scientific theories that embody terms defining mental entities and processes in physicalist language, those theories can be understood in antirealist terms.[48]

For example, some form of operationalism may be helpful. For instance, the claim that a certain pain is to be identified with such-and-such brain state could be understood such that when one has a pain, the scientist should be able to detect some brain state that caused or resulted from that pain (and, perhaps, is always correlated with it). In

47. Moreland, *Scaling the Secular City*, pp. 77–103.

48. An alternative to my antirealist strategy would be what is called the complementarity view of integrating science and theology, roughly, the view that science and theology describe, in realist terms, two different though complementary aspects of the same phenomenon (e.g., the physical and mental aspects of mind). I think the complementarity view is helpful and valuable as a part of an overall theory of integrating theology and science. But my antirealist strategy takes into account two important factors sometimes left out by the complementarity view. First, when a scientific definition of some phenomenon is taken as the whole of that phenomenon, the scientific definition is sometimes best viewed in antirealist terms. The success of the definition may not warrant a realist interpretation of the definition in question. Second, the scientific definition often does not attempt to describe the physical aspect of a phenomenon, but instead tries to offer testable criteria for the presence of the phenomenon in question. Thus, it seems more reasonable to say that mental events do not have two sets of properties, mental and physical, such that two levels of property description are being offered, as the complementarity view may suggest. Rather, the physicalist language about neurons firing, and so forth, may merely be offering operationally testable criteria for brain correlates of the mental events themselves. In this case, if the scientific definition is offered as a definition of pain or love, then it should be treated in antirealist terms and not as a realist definition of the physical properties of love-events in the mind/brain.

this case, mental phenomena are given empirical, testable import through procedural definitions that allow one to test their presence without falling into the trap of identifying them with the tests or with the physical phenomena whose presence they signify.

Antirealism would also be appropriate in cases where psychologists and social scientists identify normative notions in empirical, descriptive terms. For example, I once heard a psychologist lecture on what constituted a mature modern woman. When I asked her what *mature* meant, she responded by telling me that *mature* was defined by falling in a certain range of scores on certain tests for defensiveness and other traits. Prima facie, terms like *normal, duty, mature,* and *psychologically healthy,* are normative, evaluative, prescriptive moral terms. As such, their proper domain of study as normative terms is philosophy and theology. Science focuses on the descriptive, factual *is,* not the prescriptive, moral *ought.*[49]

Consider a final example the biological notion of a species as applied to being a human being. Biologists offer various definitions of what it is to be a member of a species. For example, a species may be defined as a group of interbreeding demes (a deme is a small, local population of organisms that maintains its identity) that are reproductively isolated from other groups of organisms. Or a species may be defined as a single ancestral lineage that maintains its biological identity from other such lineages and has its own evolutionary tendencies and historical fate.

These definitions of species use notions that can be given more precise, testable content: reproductive isolation, deme, biological identity, evolutionary tendency, and so on. Now, there is a philosophical or theological notion of a species that differs from the biological one. Philosophers discuss issues about natural kinds of things: what is it that all members of the natural kind *human being* have in common, what unites into a unity the properties, parts, and dispositions of a particular human being, how is it that a human can undergo change and still remain the same thing? The theological notion of a human being involved attempts to specify the image of God, and so forth. While not necessarily incompatible, the philosophical and theological notion of the natural kind *human being* is different from the biological species *homo sapiens.* Any attempt to equate the two leads to difficulties.

For example, in a recent book, philosopher James Rachels argues

---

49. I have discussed matters about ethical naturalism and nonnaturalism elsewhere. See *Scaling the Secular City,* pp. 108–13. See also Basil Mitchell, *Morality: Secular and Religious* (Oxford: Clarendon, 1980), pp. 1–29.

that active euthanasia (actively killing a human being) is morally permissible if a person no longer has anything to live for from his or her own point of view.[50] Part of his argument centers on defending the idea that a person does not have moral value just because he is a human being (i.e., a member of the biological species *homo sapiens*). Rachels fails to see that being a human being is not simply a matter of falling under a conception of species biologically defined. A human being is one who falls under a natural kind concept that includes having moral value or dignity among the properties constitutive of being human.[51] Rachels sees a scientific definition of a human being as stating the whole of what it is to be human.[52]

In all these cases, an allegedly successful scientific theory describes some phenomenon, or conditions associated with some phenomenon, in reductive terms. The phenomenon in question appears to lie properly outside the bounds of science but is, nevertheless, exhaustively defined in scientific terms. When such cases arise, I suggest that the success of the scientific theory be understood in antirealist terms. The theory may merely save the phenomena, that is, accurately describe and predict the observational phenomena associated with mind, value, humanness, and the like. It may give empirically testable, operational definitions of the thing in question that allow us to discover scientific correlates, causal relations, or properties of the thing under investigation.

In cases like this, the success of the scientific theory should not draw us into reductionism. Neither should it lead us to think that the scientific description really describes the world and the philosophical or theological description merely offers a language game, a way of looking at the phenomenon in meaningful ways. Exactly the opposite may be true.

Before moving on to the second criterion, I should point out that the first uses the terms *appropriate* and *inappropriate*. These may appear to be weasel words, vague and hard to specify precisely in a non-question-begging way. My only defense is to reiterate what has already

50. See James Rachels, *The End of Life* (Oxford: Oxford University Press, 1980). For a critique of Rachels, see J. P. Moreland, "James Rachels and the Active Euthanasia Debate," *Journal of the Evangelical Theological Society* 31 (March 1988): 81–90.

51. See David Wiggins, *Sameness and Substance* (Cambridge: Harvard University Press, 1980), pp. 149–89.

52. For more on the nature of being human and the biological notion of species, see Etienne Gilson, *From Aristotle to Darwin and Back Again: A Journey in Final Causality, Species, and Evolution* (Notre Dame: University of Notre Dame Press, 1984); Arthur Peacocke, *God and the New Biology* (San Francisco: Harper and Row, 1986); Sober, ed., *Conceptual Issues in Evolutionary Biology*, pp. 529–702; several articles in volume 51 (1984) of *Philosophy of Science*.

been said: the application of these terms most likely cannot be settled in a general way, but must be argued for on a case-by-case basis.

An antirealist approach should be taken toward some scientific theory in those cases where a realist view conflicts with a rationally well-established conceptual problem for that theory but an antirealist view does not. The point of this criterion is twofold. First, it allows for the fact that science is not an airtight approach to or set of disciplines about the world that is independent of other disciplines. By presenting external conceptual problems, other disciplines like philosophy and theology can appropriately enter into the evaluation process regarding some scientific theory. The point is to make science consistent with other putative knowledge and to show that, when conflict arises between science and another field, it is not just the other field that should be adjusted. It may be that the scientific theory in question should be modified, and one way to do this is to interpret it in antirealist terms.

Second, the criterion allows for the fact that scientific theories will sometimes have internal conceptual problems, and an antirealist interpretation of the theory may resolve the problem. For example, some have argued that the wave/particle nature of light is conceptually unclear or even contradictory if understood in realist terms. If this is a serious internal conceptual problem, then we could respond by treating the wave/particle duality in antirealist terms.

Again, some interpretations of quantum phenomena have led scientists and philosophers to abandon classical logic for various forms of what are called deviant logic. (Classical logic accepts the law of excluded middle [A or not-A] and the law of noncontradiction [something cannot be A and not-A at the same time in the same way]. Deviant logics abandon these and postulate many-valued logics, e.g., a statement is either true, false, or indeterminate.) For example, in the "two-slit" experiment, where light is fired at a photographic plate through a dark screen with two holes, an interference pattern arises. Some explain this in terms of seeing light as a series of photons where a single photon is considered to be in two states at once, breaking through both holes. Heisenberg's uncertainty principle is often understood to imply that a particle like an electron does not have a simultaneous, determinate set of positions and momentums. Thus, a statement that electron b has such-and-such a position can be true, false, or indeterminate. One way to resolve these internal conceptual problems is to take an antirealist approach to the quantum phenomena in question (e.g., the indeterminacy in nature is due to the uncertainty of our knowledge of nature or our inability to measure nature and is not a feature of nature itself).

External conceptual problems can be treated in the same way. I have argued elsewhere that good philosophical arguments can count against any model of the universe that postulates an actually infinite, beginningless past.[53] Suppose, for the sake of argument, that these philosophical arguments are reasonable. Now suppose that some scientific model arises in the near future that is fairly successful at explaining, describing, and predicting observational phenomena and entails a beginningless universe. Any such model would have these philosophical arguments as external conceptual problems and they would tend to count against the rational acceptance of that model. One way to relieve this cognitive tension would be to interpret the successful scientific theory in antirealist terms.

As with the first criterion, the second contains what may appear to be a weasel phrase, namely "a rationally well-established conceptual problem." When is a conceptual problem rationally well established, and how do we weigh the epistemic tradeoffs involved in adopting an antirealist view of the scientific theory in question versus adopting a realist view and dismissing or suspending judgment about the conceptual problem? Again, a case-by-case study is needed to solve this problem and to surface more precise criteria for answering these questions.

Other criteria could be suggested for adopting an antirealist approach to some scientific theory: the field of science is relatively young; the history of theories in the domain under investigation has exhibited a large proportion of theory replacements versus refinements; the field is currently in a Kuhnian period of crisis; nonrational, sociological factors account for much of the theory's acceptance by the scientific community; the main virtue of the theory is its empirical adequacy, and its more metaphysical, theoretical aspects can be understood as unnecessary, excess baggage, and so on.

My main point is that an eclectic model of science is appropriate and can be used in the integration of science and theology (and philosophy), and criteria can and should be developed that, in conjunction with specific case studies, would help the Christian community in its integrative endeavors. We should not automatically assume that if science and theology conflict, then either theology should be readjusted or the scientific theory should be read in realist terms and attacked accordingly.

Conclusion

This chapter has surveyed various alternatives to rational realism and noted the value of sensitivity to antirealist alternatives in discus-

---

53. See Moreland, *Scaling the Secular City*, pp. 15–42.

sions of science and theology. We have also looked briefly at an outline of an eclectic approach to science. In this regard, it seems fitting to close with some advice Gordon Clark gave a quarter of a century ago to a theologian who would try to integrate his or her theology with science:

> . . . the theologian must have an overall view of science as a whole. He must have a philosophy of science; that is, he must know what science is . . . any argument for or against religion, any argument that claims scientific support, depends more on the philosophical implications of science than on bits of detailed information. One must fit his science into a general philosophy. One must consider the range and the limitations of scientific application. One must know what science really is.[54]

54. Clark, *Philosophy of Science*, p. 8.

The Scientific Status of Creationism

And, I learnt what a hollow sham modern day Creationism really is: crude, dogmatic, Biblical literalism, masquerading as genuine science. . . .
> —Michael Ruse, in *Philosophy of Science* (1984), reviewing *Abusing Science*

A more proper approach is to think in terms of two scientific models, the *evolution model* and the *creation model*.
> —Henry Morris, *Scientific Creationism*

It may be time, in other words, to repair the breach that has opened up between the Darwins and the Paleys, to acknowledge that they were never that far apart, and to continue searching for a conception of the origin, end, and purpose of life that invites not only our continuing study but also our praise.
> —Charles Henderson, Jr., *God and Science*

Most people in the United States cannot help being aware of the recent debates about creation and evolution, especially the aspect of those debates that has centered on the scientific status of creationism, also called creation science or scientific creationism. Is creation science a contradiction in terms? Is it really a religion and not a science at all? Far more heat than light has been generated around this issue, with emotionalism, ad hominem remarks, and a good deal of misunderstanding tainting both sides of the dispute.

This chapter investigates whether creation science is real science or just disguised religion. Three preliminary remarks are in order. First, we will not focus on whether the scientific evidence favors creationism or evolution. That is a first-order scientific question that involves, among other things, an assessment of prebiotic soup scenarios of the origin of life, the nature of the fossil record, limits to evolutionary

213

change, and so forth.[1] The question before us is the second-order philosophical question of whether scientific creationism should be classified as science at all.

Second, we will not debate the proper way to integrate science and theology or the exegetical issues involved in interpreting the early chapters of Genesis. I have discussed these issues elsewhere.[2] More to the point, a number of Christian thinkers believe both that a proper understanding of science and theology involves the possibility of direct interaction and potential conflict between theological and scientific assertions and that statements about the world can be derived from Holy Scripture and tested scientifically. I happen to be among them, but that is not immediately relevant. For the sake of argument, let us assume this model of interaction between theology and science, along with the concomitant family of interpretations of Genesis associated with it. Our question will be whether this approach somehow violates the nature of science.

Finally, we have already learned a number of lessons in chapters 1–5 that imply that creation science is, in fact, a science (e.g., problems of defining the necessary and sufficient conditions for some activity to count as science, the role of theology and philosophy in the history of science, the existence of external conceptual problems as part of science, and so forth). We will not rehearse those lessons here, except as they bear on specific issues relating to the creation science controversy. Instead, we want to focus on some of the details of the controversy itself.

Historical Background

The current debate about creationism should be seen in light of its historical context.[3] Evolutionary ideas have been around for a long

1. See the bibliography for a list of some of the more important works focusing on the scientific evidences involved in the creation/evolution debate.

2. See J. P. Moreland, *Scaling the Secular City: A Defense of Christianity* (Grand Rapids: Baker, 1987), pp. 200–220.

3. For brief overviews of this historical context see Charles Henderson, Jr., *God and Science* (Atlanta: John Knox, 1986), pp. 42–62; Michael Denton, *Evolution: A Theory in Crisis* (London: Burnett Books, 1985), pp. 17–78; Colin Russell, *Cross-Currents: Interactions Between Science and Faith* (Grand Rapids: Eerdmans, 1985), pp. 127–76; I. Bernard Cohen, *Revolution in Science* (Cambridge, Mass.: Belknap, 1985), pp. 283–300; Norman L. Geisler and J. Kerby Anderson, *Origin Science* (Grand Rapids: Baker, 1987), pp. 71–87; J. K. S. Reid, *Christian Apologetics* (Grand Rapids: Eerdmans, 1969), pp. 167–81. More detailed studies can be found in William R. Coleman, *Biology in the Nineteenth Century: Problems of Form, Function, and Transformation* (Cambridge: Cam-

time. The Greek philosopher Anaximander (610–546 B.C.) believed that living things originated by natural processes from moisture and were originally aquatic beings that later transferred to dry land. However, evolutionary ideas never gained great currency for a number of reasons: the apparant teleology or design in the universe; the stable coadaptation of parts in an organism that would be disrupted by the minor changes hypothesized in transitional forms; the fixity or stability of species derived from Platonic, but more specifically Aristotelean, essentialism (each member of a species has the same, unchanging essence); the possibility of a neat, orderly taxonomic system of organisms, a system that would be difficult if numerous transitional forms existed; and the simple observation that like begets like. Thus, even up to the 1850s and beyond, many men of science believed it was reasonable to hold a discontinuous, typological view of nature and to believe that the Bible recorded the actual history of the creation of life and thus could serve as a guide for doing science.

When Charles Darwin's *Origin of Species* appeared in 1859, it caused a major revolution in science itself. The primary debate between creationists and evolutionists was not over fossils, biological classification, or geological evidence, though these were involved. The major debate that Darwin surfaced was between two competing epistemes. As Neal Gillespie puts it, "... Darwin's rejection of special creation was part of the transformation of biology into a positive science, one committed to thoroughly naturalistic explanations based on material causes and the uniformity of nature. . . ."[4]

An episteme is a set of common assumptions about the nature and limits of knowledge. It is roughly synonymous with a paradigm in Kuhn's sense of a world view that includes a picture of what counts as good science. An episteme involved the idea of science itself: what counts as good science, how science should be practiced, the limits of science, and so on. Thus, a scientific episteme is not just a view within science about the nature of living organisms and their development. It is also a second-order philosophical view about science that defines the nature, limits, metaphysics, and epistemology of "good" science.

---

bridge University Press, 1971); John C. Greene, *Science, Ideology, and World View: Essays in the History of Evolutionary Ideas* (Berkeley: University of California Press, 1981); Neal C. Gillespie, *Charles Darwin and the Problems of Creation* (Chicago: University of Chicago Press, 1979); Barry Gale, *Evolution Without Evidence: Charles Darwin and the Origin of Species* (Albuquerque: University of New Mexico Press, 1982); David Lindberg and Ronald Numbers, eds. *God and Nature: Historical Essays in the Encounter Between Christianity and Science* (Berkeley: University of California Press, 1986).

4. Gillespie, *Charles Darwin*, p. 19.

During the Darwinian revolution, two main epistemes competed with each other: creationism and positivism.[5] Creationism was not just a view about how life began. More importantly, it was an episteme on a par with positivism. It was a view about how science could and should be practiced. The creationist saw the world and everything within it as the result of direct or indirect divine activity. His science was inseparable from his theology. An understanding of nature was an understanding of the divine mind, it was thinking God's thoughts after him. The creationist believed that science was logically and theoretically obligated to theology and that it was legitimate to consult the early chapters of Genesis as a guide for biology and geology. Creationism saw design in the universe; the stability of species; the reality of natural, causal laws grounded in the command of God; and the use of final causes or teleological explanation and vital influences, all as parts of science.

Positivism was the rival episteme—not the empiricism or phenomenalism we discussed in chapter 5, but a naturalistic view of science itself. Its point was to get theology out of biology and let biology be autonomous, like physics and chemistry. Positivism generally rejected teleology, vital forces, and the supernatural within science and sought to explain biological phenomena in naturalistic terms emphasizing material and efficient, often mechanistic, causality. Positivists were skeptical about the role of theology in science. Thus, Darwin's theory signaled the epistemic breakdown of theology as a vehicle for doing science.

Creationists gave support for their ideas. In addition to some of the arguments listed a few paragraphs above, creationists predicted limits to breeding, an absence of transitional forms in the fossil record, the inadequacy of a purely physical explanation of life, the existence of evidence for a worldwide catastrophe (Noah's flood), and so on. But positivists believed that creationism was not a fruitful guide for biological research. In particular, it left out of the domain of science (as defined by positivism) questions about the origin of first life and species. Darwin's theory gave a scientific view of the origin of species within a purely naturalistic framework.

One other issue in the creation/evolution debate should be mentioned. Evolutionary theory had far-reaching implications for man's view of the world, his own identity, and the nature of religion. As Michael Denton has observed:

The triumph of evolution meant the end of the traditional belief in the world as a purposeful created order—the so-called teleological outlook

5. Ibid., pp. 1–66.

which had been predominant in the western world for two millennia. According to Darwin, all the design, order and complexity of life and the eerie purposefulness of living systems were the result of a simple blind random process—natural selection. Before Darwin, men had believed a providential intelligence had imposed its mysterious design upon nature, but now chance ruled supreme. God's will was replaced by the capriciousness of a roulette wheel. The break with the past was complete.[6]

Darwinianism added to a naturalistic outlook on the world a view of man as a modified monkey who had no ultimate meaning and purpose in life and a view of religion as a matter of private faith and practice separate from science. I am not claiming that evolutionist thinkers since Darwin have not attempted to maintain the dignity of man, the meaning of life, or even the truth of Christianity in light of evolution. Advocates of the complementarity model of science and theology come readily to mind. But evolutionary theory did have the impact Denton suggests, even if he overstates it, and thinkers have labored within Darwinism, as men felt no need to labor before Darwinism, to maintain dignity and meaning for man and the cosmos. It should be fairly obvious why those philosophical/theological ideas are more problematic after Darwin than before him. These ideas are merely consistent with (or complementary to) evolutionary and physicalist visions of man. They are not rational implications that naturally flow from Darwinism.They are more at home in the creationist world view. That is one reason why advocates of theistic evolution utilize fideism to "support" their religious views.

In sum, the creation/evolution debate of the nineteenth century was not simply about scientific evidence or facts.[7] More importantly, it was a philosophical debate between two epistemes regarding the issue of what science was and how it should be practiced. It was also a debate about the broad cultural implications of evolution. Thus, the evolution/creation debate in the nineteenth century and, I would argue, up to now, has been a largely philosophical debate about how to view science, theology, man, morality, and the cosmos. Once we see the philosophical nature of the creation/evolution debate, we can understand why the disputants often are emotional.

There is a lesson to be learned from this brief historical sketch. When we evaluate alternatives in creation and evolution, it is not without precedent to bring philosophical, theological, and other considerations to bear. In light of earlier chapters, where we have argued that science is not totally removed from other fields of study, it is

6. Denton, *Evolution*, p. 15.
7. For a good summary of the scientific evidences involved with the acceptance of Darwinianism, see Gale, *Evolution Without Evidence*, pp. 165–69.

appropriate to bring rational problems from other disciplines to bear on the creation/evolution debate.

Darwin himself did this frequently. For example, he often mused about how he could trust the deliverances of the human mind (especially its scientific theorizing, his own system included), if the human intellect were the result of a blind process of irrational causes.[8] In so doing, Darwin was raising an external conceptual problem, a philosophical issue in this case, against evolutionary theory. In our day, philosopher of science Stanley Jaki has raised the same difficulty for a purely naturalistic evolutionary theory.[9] The point here is not to investigate the seriousness of this problem. Rather, historical precedent and philosophy of science unite in reminding us that such issues are relevant to the creation/evolution controversy.

What Is Scientific Creationism?

In chapter 1 we saw how difficult it is to give a set of necessary and sufficient conditions for something to count as science. The same problems arise in defining scientific creationism. For our purposes, we can offer a working characterization of creation science that is adequate as long as we realize that it is not intended to be an airtight definition. Our main goal is to spell out creation science with sufficient clarity to see if it can overcome accusations that it is not really science.

In December 1981, a creation/evolution trial (*McLean* v. *Arkansas*) was held in Little Rock, Arkansas.[10] The following definition of creation science was given:

8. Cf. *Metaphysics, Materialism, and the Evolution of Mind: Early Writings of Charles Darwin*, transcribed and annotated by Paul H. Barrett (Chicago: University of Chicago Press, 1974).

9. See Stanley L. Jaki, *Angels, Apes, and Men* (La Salle, Ill.: Sherwood Sugden, 1983), pp. 51–72. See also George Schlesinger, *Religion and Scientific Method* (Dordrecht, Holland: D. Reidel, 1977), pp. 83–84. Philosophers have responded by trying to offer what is called an evolutionary epistemology. See Christopher Hookway, ed., *Minds, Machines, and Evolution* (Cambridge: Cambridge University Press, 1984). Evolutionary epistemology is a species of a more general approach to epistemology, naturalized epistemology, that tries to place the normative study of epistemology within the descriptive domain of psychology and claims that one cannot study how we ought to arrive at our beliefs independently from how we do, in fact, arrive at them. For more on this, see Kornblith, ed., *Naturalizing Epistemology* (Cambridge, Mass.: MIT Press, 1985). Similar issues have arisen in the area of evolutionary theory and morality. See Jeffrie G. Murphy, *Evolution, Morality, and the Meaning of Life* (Totowa, N.J.: Rowman and Littlefield, 1982).

10. For coverage of the trial, see Norman L. Geisler, *The Creator in the Courtroom* (Milford, Mich.: Mott Media, 1982).

Creation-science means the scientific evidences for creation and inferences from those scientific evidences. Creation-science includes the scientific evidences and related inferences that indicate: (1) Sudden creation
of the universe, energy, and life from nothing; (2) The insufficiency of
mutation and natural selection in bringing about the development of all
living kinds from a single organism; (3) Changes only within fixed limits
of originally created kinds of plants and animals; (4) Separate ancestry
for man and apes; (5) Explanation of the earth's geology by catastro-
phism, including the occurrence of a worldwide flood; and (6) A relatively recent inception of the earth and living kinds.

Before we consider criticisms of the scientific status of creation science, some comments about this definition are in order. First, scholars
have offered several different interpretations of the early chapters of
Genesis, several of which are embraced by evangelicals.[11] These interpretations have given rise to three broad schools of thought about
evolution, each with different varieties: theistic evolution, progressive
creationism, and special creationism.

Theistic evolution holds, roughly, that evolutionary theory is approximately true, and the how and what questions of the origin and
development of life are solely within the province of science. Theology
tells us who guided the process by using the mediate operation of
secondary causes and why God brought life and man into existence.
It does not tell us how or when. So understood, theistic evolution has
not entered the debate about creation/evolution except when the debate centers on naturalistic versus theistic evolution. Most theistic
evolutionists have embraced the positivist episteme discussed earlier
and thus regard creation science as unscientific and the creationist
episteme as an inappropriate view of science.

Progressive creationism comes in several versions, but its center
involves these theses: First, the cosmos is old (15–20 billion years), the
earth is old (4–4.5 billion years), and life is old (3.98 billion years?).
Progressive creationists are divided about the antiquity of man, but
many hold that Adam and Eve, the original pair of human beings, are
recent in appearance, within 20–50,000 years. Second, progressive creationists differ about the extent of Noah's flood, some holding to a
worldwide flood and others to a local flood covering the entire contemporary human race. But they agree that Noah's flood played little
or no role in forming the geologic column. Third, progressive creationists agree that the general theory of macroevolution is false and
that God directly, immediately, and with primary causality created
the first life, the general created kinds of life, and man. They differ
over when these various acts of creation took place but generally hold

11. See Moreland, *Scaling the Secular City*, pp. 214–20.

the days of Genesis to be long periods of time. Advocates of progressive creationism are Walter Bradley, Charles Thaxton, Roger Olsen, Robert Newman, and Herman Eckelmann.[12]

Special creationists include Henry Morris, Duane Gish, Gary Parker, and A. E. Wilder-Smith.[13] We need not discuss their view here since the definition given in the Arkansas trial reflects it.

A number of evangelicals who hold to the inerrancy of Scripture believe either that the Bible does not teach a universal flood or that, if it does, such a flood was not the main factor shaping the geologic column.[14] The same can be said about the biblical dating of the cosmos, earth, and living kinds, with the exception of human beings. Thus, progressive creationists and special creationists agree about points 1–4 of the Arkansas definition of creation science but differ on points 5 and 6.

In my opinion, points 1–4 are the strongest in light of current scientific data, but such an opinion is not our present concern. Our concern is whether creation science is science at all. Thus, I will defend the scientific status of special creation, but that should not be taken to imply that I do not consider progressive creationism a legitimate option. In my view both progressive creationism and special creation have merits and problems, and the Christian community is enhanced by the coexistence of these models. We should encourage continued dialogue and the development of each model.

Is creation science really religion masquerading as genuine science, or is it science? In the last section of this chapter we will examine some of the positive scientific features of creation science and try to see what philosophical, epistemic values are at stake in the creation/evolution controversy. For now, let us defend the scientific status of creation science by responding to objections to it.[15]

12. See Charles Thaxton, Walter Bradley, and Roger Olsen, *The Mystery of Life's Origin: Reassessing Current Theories* (New York: Philosophical Library, 1984); Robert C. Newman and Herman J. Eckelmann, *Genesis One and the Origin of the Earth* (Downers Grove: Inter-Varsity, 1977). See also John Weister, *The Genesis Connection* (Nashville: Thomas Nelson, 1983).

13. Henry Morris, *Scientific Creationism; Creation and the Modern Christian* (El Cajon, Calif.: Master, 1985); Henry Morris and Gary Parker, *What Is Creation Science?* (San Diego, Calif.: Creation-Life Publishers, 1982); Duane Gish, *Evolution: The Challenge of the Fossil Record* (El Cajon, Calif.: Master, 1985); A. E. Wilder-Smith, *Man's Origin, Man's Destiny* (Wheaton: Harold Shaw, 1968).

14. See Bernard Ramm, *The Christian View of Science and Scripture* (Grand Rapids: Eerdmans, 1954), pp. 156–69. For a criticism of this view, see Morris, *Scientific Creationism*, pp. 250–55.

15. Cf. Norman L. Geisler, "Is Creation-Science Science or Religion?" *Journal of the American Scientific Affiliation* (September 1984): 149–55; Davis Young, "Is 'Creation-Science' Science or Religion?—A Response," *Journal of the American Scientific*

Objections to Scientific Creationism's Scientific Status

*Creation science uses religious concepts ("God," "creation," "kind");
therefore it is religion, not science.* This objection is frequently encountered in popular news media and was explicitly raised by Judge Overton in the Arkansas trial. Said Overton, ". . . both the concepts and
wording of [the creation science definition] convey an inescapable re-
ligiosity. Section 4(a)(1) describes 'sudden creation of the universe,
energy and life from nothing.' Every theologian who testified, including defense witnesses, expressed the opinion that the statement referred to a supernatural creation which was performed by God."[16]

This objection fails because *God* (like the other terms) is not necessarily a religious concept. When *God* functions as a religious concept, it is used to promote religion as its principal or primary effect,
it is involved in moral and spiritual exhortation, and it is surrounded
by ritual and other forms of religious devotion. But the term *God* may
be a mere philosophical concept or theoretical term denoting an explanatory theoretical entity needed in some sort of explanation, much
like the terms *quark* and *continental plate.* For example, in Aristotle's
view of the world, *God* referred to a theoretical entity that served to
explain, among other things, motion in the universe. But Aristotle did
not worship the unmoved Mover of his philosophical system, and discussion of his notions of God in a philosophy class are not religion.
Isaac Newton appealed to the existence of God in order to help explain
certain irregularities in planetary motion, arguing that God occasionally nudged the planets. Newton's appeal was later falsified, but that
is not the point. The point is that God was not a religious concept here
but a theoretical one, for God was not the object of worship or a means
of moral exhortation in Newton's theory but a mere explanatory entity.

The notions of a creator and creation are not necessarily religious
either. It is generally agreed that these notions do not constitute the
essence of religion, for several religions (Theravada Buddhism, various

Affiliation (September 1984): 156–58; Larry Laudan, "Commentary: Science at the
Bar—Causes for Concern," *Science, Technology, and Human Values* 7 (Fall 1982): 16–
19; Michael Ruse, "Response to the Commentary: *Pro Judice*," *Science, Technology,
and Human Values* 7 (Fall 1982): 19–23; Larry Laudan, "More on Creationism," *Science, Technology, and Human Values* 8 (Winter: 1983): 36–38; Philip Quinn, "The Philosopher of Science as Expert Witness," in *Science and Reality*, ed. James T. Cushing,
C. F. Delaney, and Gary Gutting (Notre Dame: University of Notre Dame Press, 1984),
pp. 32–53; Larry Laudan, "The Demise of the Demarcation Problem," in *Physics, Philosophy and Psychoanalysis*, ed. R. S. Cohen and L. Laudan (Dordrecht, Holland:
D. Reidel, 1983), pp. 111–128.
  16. See Geisler, *Creator in the Courtroom*, p. 173.

forms of Hinduism) have no notion of creation. *Creator* can mean "a proximate or ultimate first cause." One could add that the first cause is personal in the sense that it has intentions and acts in light of knowledge and purposes.[17] The notion of a person can be scientific and is used in psychology and sociology. And behavior, even intentional behavior, is a biological notion often ascribed to animals. Thus, the notion of a personal first cause does not seem to be unscientific.

Similar observations could be made about the terms *creation* and *kinds*. *Creation* simply refers to something coming to be. Scientists refer to the first event of the Big Bang as a creation in this sense, and advocates of the now-abandoned steady state theory of the universe postulated the continuous creation of hydrogen. These uses of *creation* are not religious, and neither is the one in the definition of creation science.

As for *kind*, University of California at Davis geneticist Francisco Ayala claimed that it is a religious, not a scientific, term.[18] But I fail to see what Ayala means here. If he means that the term is religious because it is used in the Bible, then so are a lot of other terms used by science—mathematical numbers, cattle, trees, the sun, and so forth. A better objection to *kind* is that it is vague: it is used in Scripture for various levels of classification from subspecies to phylum, and creation scientists have not given precise operational definitions for its use. There is a point to this objection, but its real point is not that *kind* is religious and not scientific but that it is a vague scientific term that needs to be made more precise. After all, one of the developmental aspects of science is the increased clarity and precision given to its terms. But when a term is made more precise, the term does not change from nonscientific to scientific, it changes from vague to precise.

In sum, none of the terms used in the creation science definition are religious. They may be used in the Bible, but they have been transformed into general philosophical and scientific terms that, in turn, can be given more and more precise operational language. Thus, this objection fails.

God (creation, kind) *is an illegitimate term in science, not because it is religious, but because it is supernatural, and science explains by using natural laws.* Michael Ruse provides an example of this criticism:

17. For a helpful treatment of God as agent, see William P. Alston, "God's Action in the World," in *Evolution and Creation*, ed. Ernan McMullin (Notre Dame: University of Notre Dame Press, 1985), pp. 197–220. Neal Gillespie offers a good overview of four different uses of "creation by law" involved in the mid-nineteenth century, including different views about God as an agent. See Gillespie, *Charles Darwin and the Problem of Creation*, pp. 22–40, especially pp. 22–25.

18. See Geisler, *Creator in the Courtroom*, p. 81.

Furthermore, even if Scientific Creationism were totally successful in making its case as science, it would not yield in a *scientific* explanation of origins. Rather, at most, it could prove that science shows that there can be *no* scientific explanation of origins. The Creationists believe the world started miraculously. But miracles lie outside of science, which by definition deals only with the natural, the repeatable, that which is governed by law.[19]

In chapter 1, we investigated the claim that science deals only with the natural, the repeatable, and what is governed by law. I will not repeat that discussion here. Instead, three brief comments are in order. First, this position begs the question in favor of a naturalistic explanation of origins, since it claims that a proper explanation must be naturalistic and that science by definition cannot recognize the intervention of a personal first cause. Yet even an atheist would agree that, in the absence of a case for contradictions among God's attributes, the existence of God is logically possible. God could exist and he could have created the universe, various kinds of life and man in particular through immediate, primary causality. If science could not, in principle, recognize this possibility because of its naturalistic assumptions, then science would necessarily lead to a false conclusion if creationism is true.

If Ruse wants to define *science* in naturalistic terms, then we can define a new term, *creascience*, that allows for the recognition of discontinuities in nature that indicate the intentional, immediate intervention of a first cause that resembles a person. Note, if God does not exist, or if he has never intervened in the world through primary causality, then *science* and *creascience* are empirically equivalent and equally adequate approaches to the study of nature. The main difference between *science* and *creascience* is that the latter allows for the possibility that primary causality has occurred and can be recognized. It therefore better satisfies the epistemic value of being able to account for a wider variety of actual or potential phenomena. Why, then, should we prefer *science* to *creascience*?

After all, Darwin himself believed that radical jumps in nature do not occur: *natura non facit saltum*, that is, "nature does not proceed in jumps." This is one reason why Darwin postulated that evolutionary development proceeds in a series of very small steps. For Darwin, if gaps prevailed in the fossil record, if evolutionary change took place in large-scale jumps, then this would be evidence of a miracle. One reason for this is that organisms are like sentences—they are isolated

19. Michael Ruse, *Darwinism Defended* (Reading, Mass.: Addison-Wesley, 1982), p. 322.

unities of coadapted parts. If a computer screen showed one sentence and then another that was totally different, this would be evidence of a mind behind the sentences. It would be unlikely that such a jump could have landed on a second meaningful sentence by chance. Similarly, if one organism is replaced by a second one that is fairly distant from it morphologically, then for Darwin and other scientists of his day, this would be evidence of intelligent intervention. In fact, several scientists during Darwin's time gave testable implications for God's immediate activity.[20]

The point here is not that all the creation-science predictions of Darwin's day were verified or that radical discontinuities in evolutionary development are currently explained by postulating God as the cause of such discontinuities. (Punctuated equilibrium theory comes readily to mind as a proposed naturalistic explanation.) The point is that Darwin and other scientists of his day were practicing science when they attempted to specify tests for the presence of personal intervention in the development of life.

This leads to a second point. Scientists of other generations recognized that God was a legitimate actual or hypothetical source of explanation in science (and some do today, including some who are not theists). Why, then, should we accept a definition of science that arbitrarily rules out as nonscience all the cases in the history of science where God was appealed to as a theoretical entity? Even Philip Kitcher, no friend of creationism, agrees:

> Moreover, *variants* of Creationism were supported by a number of eminent nineteenth-century scientists—William Buckland, Adam Sedgwick, and Louis Agassiz, for example. These Creationists trusted that their theories would accord with the Bible, interpreted in what they saw as a correct way. However, that fact does not affect the scientific status of those theories. Even postulating an unobserved Creator need be no more unscientific than postulating unobservable particles. What matters is the character of the proposals and the ways in which they are articulated and defended. The great scientific Creationists of the eighteenth and nineteenth centuries offered problem-solving strategies for many of the questions addressed by evolutionary theory.[21]

Kitcher goes on to argue that those creationist theories were, in fact, falsified, but the point here is that current scientific creationists believe their theories have not been falsified and do explain scientific

---

20. See Gillespie, *Charles Darwin and the Problem of Creation*, p. 7.

21. Philip Kitcher, *Abusing Science: The Case Against Creationism* (Cambridge, Mass.: MIT Press, 1982), p. 125.

facts. Our concern is to note that, as Kitcher notes, an appeal to God is not necessarily unscientific.

Third, it is far from clear that *God* is being used as a supernatural concept in any way inappropriate to science. In this regard, Norman L. Geisler and J. Kerby Anderson have distinguished between operation science and origin science. They argue that appealing to God as a personal first cause is legitimate in the latter but not the former.[22] Operation science is an empirical approach to the world that focuses on repeatable, regularly recurring events or patterns in nature (e.g., chemical reactions or the relationship between current, voltage, and resistance in a circuit). Operation science tests theories against these recurring patterns of events and, theologically speaking, secondary causes are the only focus.[23] Secondary causation refers to God's acting mediately (i.e., through the instrumentality of natural laws); primary causation refers to God's acting immediately (i.e., directly, such that discontinuities obtain in the world).

In contrast to operation science, origin science focuses on past singularities that are not repeatable (e.g., the origins of the universe, life, various life forms, and mankind). Such singularities can have a personal first cause, and it is within the domain of origin science to look for such causes. *God*, as a term in origin science, means, roughly, a first cause of some discontinuity or singularity who acts with intentionality in light of knowledge and purpose.

Carl Sagan has pointed out that in the search for extraterrestrial intelligence, all we need to discover is one message that contains information and not mere order, even if the message cannot be translated. In dealing with the past singularities listed, if they bore the features of intelligence (e.g., information, as in DNA, or extremely intricate, machinelike order, as in the protein synthesizing ribosome), then it would be reasonable to infer a personal cause of the singularity as a part of science itself.[24]

22. Cf. Geisler and Anderson, *Origin Science*, pp. 13–36. See also Thaxton, Bradley, and Olsen, *Mystery of Life's Origin*, pp. 204–14.

23. While I generally agree with this characterization of operation science, it seems to leave out the possibility that God *could*, and actually *does*, occasionally work through primary causality in regular ways. For example, certain psychological aspects of spiritual transformation result from direct acts of God on the self when the preconditions for God to act are present (e.g., I fast or pray in certain ways or ask to be filled with God's Spirit). These acts of God are repeatable and can even be characterized in lawlike ways (e.g., fasting in certain ways produces humility of heart and spiritual courage) that could be described psychologically but seem to involve primary causal action on God's part.

24. Cf. Lane Lester, Raymond Bohlin, *The Natural Limits to Biological Change* (Grand Rapids: Zondervan, 1984), pp. 153–61; Denton, *Evolution*, pp. 326–43. For a different approach that allows scientific data to justify postulating some extraordinary cause

It would seem, then, that there is nothing wrong in principle with using *God* as a theoretical term in science.[25] But some will object,"If we allowed appealing to God anytime we don't understand something, then science itself would be impossible, for science proceeds on the assumption of natural causality." This argument is a red herring. It is true that science is not compatible with just any form of theism, particularly a theism that holds to a capricious god who intervenes so often that the contrast between primary and secondary causality is unintelligible. But Christian theism holds that secondary causality is God's usual mode and primary causality is infrequent, comparatively speaking. That is why Christianity, far from hindering the development of science, actually provided the womb for its birth and development. Armed with the primary/secondary causal distinction, Christian scientists did not abandon a search for natural (secondary) causes simply because they believed in primary causes as well. The postulation of a primary cause must be justified—it cannot be claimed willy-nilly—and the operation/origin science distinction is one way to specify how the primary/secondary causality dichotomy can be approached.

---

to explain it, e.g., God, but that takes such inferences to be philosophical and not scientific, see Michael Slote, *Reason and Scepticism* (London: George Allen and Unwin, 1970), pp. 188–215. My disagreement with Slote is that I do not think a neat line can be drawn between science and philosophy.

25. Davis Young argues that personal intentions and mental phenomena are legitimate in psychology and archaeology but not in the *natural* sciences—physics, biology, chemistry, earth science, and astronomy. See Young, "Is 'Creation-Science' Science or Religion?" p. 156. He claims that natural science is restricted to material causes. But this claim cannot be justified philosophically or historically. Philosophically, Young simply begs the question by defining these branches of science in this way. Historically, we have already seen that mid-nineteenth-century creationists and evolutionists alike recognized the possibility of science discovering God's primary causal activity. Even if we grant that God never does this, just for the sake of argument, God surely could have done it, and there is no good reason in principle why mental intentionality cannot be detected through the movement or structure of physical phenomena in the world as it can be detected through movement and patterns of the human body. Theologically, Young's overemphasis on the complementarity of science and religion— a valid emphasis as part of an overall framework, but not as the whole framework— reduces to the vanishing point God's primary causal activity. God's actions are placed in some kind of "upper story," and this comes perilously close to a deist denial of primary causal activity. It is a red herring to claim that those who want to defend the actuality and recognizability of primary causality do not have a large enough picture of God's creative activity because they only see God acting in the gaps, i.e., in primary causal ways. God acts in both primary and secondary causal ways. After all, there was something special about what God did while the Red Sea was being parted that was not present as part of his causal activity regarding that sea before and after Israel crossed it.

In any case, neither in principle nor in actual history has the recognition of primary causes within science undercut the possibility of
science.

Creation science makes no predictions and is not empirically testable.
Douglas Futuyma has argued that creation science advocates "apply
to both evolution and creation the scientific-sounding word 'model' in
order to conceal the fact that evolution is a testable hypothesis, whereas
creation is not."[26] Elsewhere he offers this challenge: "If creationism
is science, let it make a single prediction that could show it right or
prove it wrong."[27]

Several things can be said in response to this objection. First, some
tenets of creationism are not testable in isolation from other claims
(e.g., the claim that man emerged from a direct act of God, a claim
that can only be tested in conjunction with other claims about features
of random or mindless events and features of events that result from
intelligence; postulates about probability arguments; and so on). But,
as Larry Laudan has argued, this "scarcely makes Creationism 'unscientific.' It is now widely acknowledged that many scientific claims
are not testable in isolation, but only when embedded in a larger
system of statements some of whose consequences can be submitted
to test."[28]

Furthermore, the claim that scientific creationism does not make
predictions and is incapable of empirical testing is simply false. Creationists have made predictions and retrodictions that can be empirically tested and that explain empirical facts.[29]

During the debate between the steady state and Big Bang cosmol-
ogies, several creationists predicted that data would confirm the Big
Bang model. Creationists also predicted that more and more evidence
would show that the early earth's atmosphere was an oxidizing one,
that there would be systematic gaps in the fossil record, that the fossil
record would contain no clear intermediate species, that a functional
purpose would be found for vestigial organs, that the claim that on-
togeny recapitulates phylogeny would be falsified by the discovery of
embryonic functions for the various morphological stages of embryonic development, and that porphyrins would be commonly found in
sedimentary rock.

26. Douglas Futuyma, *Science on Trial: The Case for Evolution* (New York: Pantheon, 1982), p. 179.

27. Ibid., p. 196.

28. See Laudan, "Commentary," p. 17.

29. See Morris, *Scientific Creationism*, pp. 8–13; Lester and Bohlin, *Natural Limits*,
pp. 172–75; Geisler and Anderson, *Origin Science*, pp. 181–83; David McQueen, "The
Chemistry of Oil Explained by Flood Geology," *Impact* 155 (May 1986): i–iv.

Other empirical claims made by creationists are that fossilized organisms will be systematically similar to their modern counterparts, organisms will exhibit stasis throughout their fossil history even if they are now extinct, and there will be genetic limits to biological change. Bradley and Kok predicted and confirmed that peptide bond frequencies in the polymerization of proteins would not show a nonrandom, self-ordering tendency in matter but would be statistically random.[30] In sum, it is simply false that creationists do not make predictions or empirically testable assertions.

Finally, the precision of prediction varies from one area of science to another. Predictions will tend to be more precise in physics and chemistry and less so in certain areas of biology, sociology, geology, and psychology, though these fields do contain some precise predictions, especially when they involve chemistry and physics. Furthermore, recall that scientific ideas vary from quite low-level hypotheses (e.g., Maxwell's theories of electromagnetism, Bohr's model of the atom) to more general, less testable families of theories called research programs (e.g., vitalism versus mechanism). Consider the following sequence listed in descending order of generality:

Physicalism as a world view (versus Platonism)

Physicalism in the natural world (versus vitalism)

Atomism (versus field theory)

Corpuscularianism (versus Dalton's model of the atom)

Corpuscularianism regarding the behavior of gases

A specific acid/base titration experiment

In some of the cases I have listed rivals to the view in question. As the level of generality descends, the level of falsifiability and empirical testability increases. For example, it is easier to verify or falsify the predicted results of a specific acid/base titration experiment with laboratory results than it is to verify or falsify physicalism regarding the natural world. Predictions at higher levels of generality are more vague; in fact, lower levels can be viewed as attempts to make more precise—and hence more testable—the theories at a higher level. Corpuscularianism is a way of spelling out atomism.

Now, various levels of generality are involved in different claims made by each side in the evolution/creation debate. While more work needs to be done in specifying whether some creationist (or evolution-

30. See Thaxton, Bradley, and Olsen, *The Mystery of Life's Origin*, pp. 147–50.

ary) claim is to be taken as a broad research tradition claim or a more specific, low-level hypothesis, two things seem clear: first, the relative vagueness, and even the value of prediction itself, will vary depending on the level of the claim in question; second, a broad research tradition is not unscientific merely because it has not been specified by more low-level hypotheses. Ultimately, such specification is needed. But a claim is not unscientific merely because such specification has not been reached. Either way, the request for prediction or empirical test can be naive, and its value depends on the area of science involved and the generality of the scientific claim in question. Many creationist (and evolutionary) claims are rejected because they carry vague predictions. But it may be that the claim is a broad research tradition claim and not a low-level hypothetical claim. If so, one would expect vagueness of prediction.

One final point. Creationists—special creationists and progressive creationists alike—simply must work harder at developing problem-solving strategies, predictions, empirical tests, and so forth. The foregoing remarks should not be taken to imply that such work is not important.

Creation science is a theory derived from the Bible and is therefore not scientific. This criticism is frequently encountered in popular discussions of creation and evolution (e.g., the editorial pages of newspapers). Unfortunately, it is an example of the genetic fallacy, the mistake of confusing the origin of a claim with its evidential warrant and undermining the claim by calling attention to its origin. What is relevant to the rationality of a claim is the evidence for it, not its source.

The medieval practice of alchemy was the historical source of modern chemistry, but that is hardly a good objection to the rationality of chemical theory. F. A. Kekule formulated his idea of the benzene ring by having a trancelike dream of a snake chasing its own tail in a circle. But the origin of his idea was not what mattered; what mattered was the evidential support he could muster for it.

It makes no difference whether a scientific theory comes from a dream, the Bible, or bathroom graffiti. The issue is whether independent scientific reasons are given for it. Creation scientists clearly offer reasons for creation science. Whether these reasons are adequate, as I believe many are, is not our present concern. Scientific reasons are offered; that is all creation science needs in order to count as science.

Creation scientists are narrow-minded and hold their theories so tenaciously that they are closed to a revision of their theories. Norman D. Newell has made this claim: "Finally I should like to define the word science, and explain why scientific creationism cannot be included in

its definition. Science is characterized by the willingness of an investigator to follow evidence wherever it leads."[31]

Apart from the fact that Newell has not even approximated the task
of defining science, several things are wrong with this objection. First,
even if Newell's characterization is correct, he has characterized not
science but scientists. But that is to follow a red herring and change
the subject. It is one thing to characterize science. Such a task involves
specifying the epistemology, metaphysics, process of discovery, and so
forth, of science. It is quite another thing to specify character traits
valued by, and supposedly exemplified by, scientists. The two issues
are not entirely unrelated, but they are not identical either.

Second, Newell is confused about the nature of objectivity most
central to science. Objectivity can mean at least two things relevant
to our discussion. It can mean psychological objectivity, a lack of bias
or commitment to a given outcome. Or it can mean epistemological
objectivity, the presentation of objective, public evidence and epistemic warrant in support of the rationality of a claim. One can be
psychologically subjective and biased, but epistemologically objective
(e.g., when one hopes that a certain view beats its rival but goes on to
offer rational, public evidence on behalf of his or her view). Contrary
to Newell, in science, epistemological objectivity, not psychological
objectivity, is essential. The issue is the evidence itself, not the investigator's lack of commitment to a theory or willingness to follow the
evidence.

Third, it is an example of an ad hominem fallacy to attack creation
science because of the (alleged) bias of creation scientists. What do the
(alleged) personality defects of creation scientists have to do with the
scientific status of the theory of creation science? Would it matter if
all Newtonians were Marxists? Surely this would not count against
the scientific status of Newtonian theory. In fact, creation science could
be tested, or even taught for that matter, by an evolutionist. And an
evolutionist could hardly be accused of narrow-mindedly embracing
creation science.

Fourth, many scientists in other areas of science have shown resistance to scientific change, as Kuhn and others have pointed out. In
fact, in many cases a small band of rebel scientists—like creation
scientists—reject the predominant paradigm and so help others to
overcome their bias in favor of a prevailing theory and see the world
differently.

Fifth, it is not true that creation scientists do not refine their views.

---

31. Norman D. Newell, *Creation and Evolution: Myth or Reality?* (New York: Praeger, 1985), p. xxxi.

If creationism is taken as a broad philosophical world view, then creationism has been around for a long time. But modern creationism also includes lower-level claims and shows both continuities and discontinuities with the creationism of the eighteenth and nineteenth centuries. Modern creation scientists' model differs in several respects from that of their counterparts one hundred years ago. (For instance, modern creationists do not accept the fixity of species.) So evidence does cause them to refine and review their theories.

Finally, Newell's claim is self-defeating, for it doesn't comply with its own standards. "[S]cientific creationism cannot be included in [the definition of science]," Newell wrote. Not "is not," or "should not be," but "cannot." But what if evidence led to the conclusion that scientific creationism should be included in the definition of science? What if, in other words, we found that scientific creationism used all or most of the major methods represented in the history of scientific practice? What if we learned that at least a significant proportion of creation scientists were not narrow-minded and tenacious in holding to their theories? What if we learned that many creation scientists had revised their theories in response to evidence? Evidence does, in fact, show each of these to be true.

If, according to Newell, "[s]cience is characterized by the willingness of an investigator to follow evidence wherever it leads," then he must be willing, at least hypothetically, to conclude that creation science is scientific. But he has already concluded that creation science "cannot" be science. No doubt this is precisely the narrow-minded tenacity he condemns when he thinks he finds it in creation scientists. And it leads him, just as he alleges it does them, "to be closed to a revision of [his] theory." If Newell is right in excluding creation science from science on the basis of this argument, he must exclude his own rejection of creation science from science on the same grounds.

If, then, Newell claims to be scientific in rejecting creation science, he must abandon his claim that creation science cannot be scientific, for the claim works equally against his own position.

It is true that creationism, construed as a broad research tradition, is fairly resistant to falsification and, from time to time, resorts to ad hoc explanations. But remember, that is just the way it is with broad research traditions, including evolutionism and physicalism. Theories at this level of generality involve a wide and complicated network of epistemic support, and a naive crucial-experiment view is not adequate at that level of generality. So it is no surprise, nor does it necessarily involve epistemic impropriety, that some creationists continue to hold to creationism as a broad research tradition in the face of anomalies.

Admittedly, a point is reached for any theory when the nature and number of anomalies (and the ability of its rivals to explain them) justify abandoning the theory. But it is hard to draw the line for when that point is reached, and identifying and assessing the anomalies are scientific ventures. In any case, creationists have responded to evidence and refined their low-level generalizations accordingly, even though they still believe the broad research tradition has epistemic warrant.

Creation science does not rely on positive evidence to support its case, but rather relies on problems in evolutionary theory. But just because some version of evolutionary theory is problematic, this does not mean creationism is true. Some other form of evolutionary theory may be adequate. Newell claims that "the 'proofs' of the creationists consist not of testable observations, or analysis of the basic processes of creation, but of attacks on scientists and their methods."[32] We have already seen that this objection is false in conjunction with our discussion of prediction and empirical testing. Creationists must give more detail and specificity to their claims, but it is false to say that they have not already given some. Creation science does rest on positive evidence.

Furthermore, it is not inappropriate to take evidence against one theory to be evidence in favor of a rival. This is the insight behind crucial experiments, and it involves the notion of partitioning rival theories into a set of options. If there is a small number of currently available theories at the same level of generality (so they can be compared adequately), then the failure of one hypothesis offers support to its rivals. The logical possibility of some potentially infinite set of unknown future theories emerging does not refute this point, for one must work with the major rivals at hand. Suppose one were evaluating some range of phenomena that three major theories were offered to explain, T_1–T_3. Suppose T_4 is the belief that some future unknown theory will solve the problem. If T_1 is refuted or weakened, then T_2 and T_3 are given positive support. The mere possibility that some future theory may be discovered does not change the rationality of the current theory. Rationality judgments usually are made in light of the major current rivals.

Now consider the case of the origin and development of life. There are currently three broad families of theories (with different low-level varieties of each) regarding the development of life: creationism (both special and progressive), neo-Darwinianism, and punctuated equilibrium. The neo-Darwinian version of macroevolution holds that evolutionary change comes about in a large number of sequential steps

32. Ibid., p. xxxii.

involving very small changes at each point along the way. But the fossil record simply has not offered what Darwin and the neo-Darwinists predicted. Millions of fossils have been discovered, and only a very small number of (alleged) transitional forms has been found. Neo-Darwinism predicts that thousands and thousands should be found.

Currently, three options are open to scientists in their attempt to explain the fossil record. First, one could choose to reaffirm neo-Darwinism, but add some ad hoc hypothesis explaining the absence of transitional forms. For example, neo-Darwinists sometimes argue that because transitional forms were not clearly superior in the struggle for reproductive advantage, they would not last long and, thus, would leave few fossils. Second, one could embrace a different version of evolutionary theory called punctuated equilibrium theory.[33] According to this view, evolutionary change occurs rapidly and is followed by long periods of stasis where no change occurs. Thus, the fossil record does not teem with transitional forms, for there were none. Third, one could embrace scientific creationism and argue that there are no clear transitional forms in the fossil record because there were no transitional forms in life, since God created the various kinds of life directly.

Proponents of punctuated equilibrium theory argue that the absence of a large number of transitional forms tends to falsify neo-Darwinism, for even if transitional forms were not superior in the struggle for reproductive advantage, one would still expect to see more of them in the fossil record than we find, if neo-Darwinism were true. This is a good argument. The fossil record tends to falsify neo-Darwinism, and for this and other reasons it tends to support punctuated equilibrium theory. But what is sauce for the goose is sauce for the gander, and there is no reason to hold that the fossil record does not similarly support scientific creationism because it tends to falsify neo-Darwinism.

In sum, creationists do rely on positive evidence to support their theories and it is not inappropriate to allow negative evidence against a rival to count as positive evidence for creationism.

We have considered six of the main objections to the scientific status of creation science. All of them have failed. In this regard, the following statement by philosopher Ernan McMullin about Judge Overton's ruling against the scientific status of creation science seems appropriate:

Philosophers of science were unhappy about the first of these claims [that creation science fails to exhibit a set of necessary and sufficient

33. For a helpful introductory comparison between neo-Darwinianism and punctuated equilibrium theory, see Lester and Bohlin, *Natural Limits*, pp. 65–148.

conditions that demarcate science from nonscience], since a "principle of demarcation" between science and nonscience, prompted by Popper in the '50s and '60s, failed to establish anything like a set of "essential characteristics" for science and indeed gave good reason to doubt that any such set could exist. The "essential" character of every single one of the characteristics listed by Judge Overton has, in fact, been challenged somewhere or another in the literature of the last twenty years.[34]

McMullin goes on to argue that creation science has, in fact, been tested and falsified. While I do not agree, the point here is that creation science is at least a science and not a religion, and it should be evaluated by normal canons of scientific appraisal surfaced by studying the philosophy and history of science itself.

There are, however, two further objections to the scientific status of creation science that are often expressed, not as knock-down arguments against the scientific status of creation science but as attitudes of skepticism about the appropriateness of theology for science and the lack of fruitfulness of creation science itself. These two issues warrant separate attention.[35]

Scientific Progress and the Prestige of Science

Recently I was on a radio talk show where the following problem— one I have heard quite frequently—was presented to me: Science shows a great deal of progress when compared with theology and philosophy. Because of this progress, it is appropriate to view science as a more rational discipline than the others. Science progresses in solving its problems, but philosophy and theology do not. Thus, the latter should be objects of rational suspicion, especially if they conflict with science. Creationism is tied to a theological framework. Thus, its rational appeal is weak compared to a robust scientific theory like evolution, which is not tied to a theological framework.

Let us set aside questions about what it means to be "tied to" a theological framework, as well as observations to the effect that evolution is "tied to" process philosophy and theology. I do think creation

34. Ernan McMullin, "Introduction: Evolution and Creation," in *Evolution and Creation*, ed. Ernan McMullin (Notre Dame: University of Notre Dame Press, 1985), p. 46.

35. A third issue I will not discuss here is the charge that creation science involves a problematic god-of-the-gaps strategy for integrating science and theology. I discussed this in *Scaling the Secular City*, pp. 204–8.

science is committed to giving a scientific defense of theological ideas in a way not necessarily present in evolution.

Science exhibits two kinds of progress. The first is technological. Over thirty-five years ago theologian and philosopher of science Bernard Ramm wrote:

> Then, too, the theoretical aspects of science found practical expressions that reached into every civilized hamlet. Steam engines, electricity, and chemistry were powerful and practical apologists for the scientific point of view. Inoculations, surgery under an anaesthetic, and brilliant new progress in surgery were medical marvels which preached irresistibly the gospel of science. What could theologians offer as a parallel to this? A theologian's product is a book, but so few of our population read the books of theologians. Further, the reasoned argument of a book cannot compete popularly with the practical gadgets of science.[36]

This type of progress is not directly relevant to our discussion. The mere presence of scientific gadgets and technological artifacts may give one a sense of awe or emotional feelings of respect, but in the absence of arguments connecting such gadgets and artifacts to the rational authority of science, this type of progress is not of concern for the epistemic authority of science vis-á-vis other disciplines.

The second type of progress is more relevant. This is the realist view that science progressively solves its problems by a fairly continual refinement of its theories toward a true description of the world.[37] Seen in this way, science shows itself to be rationally superior to theology and philosophy because science, in contrast to theology and philosophy, does not get mired in stagnant and fruitless speculations. Science makes progress in solving its problems and getting at the truth.

Several things can be said in response. First, in the absence of a successful internalist philosophy of science (where science justifies itself and, in fact, philosophy is merely an aspect of science), the rational authority of science rests on philosophy. In this regard, it is worth recalling the remarks of John Kekes:

> The scientific way of solving problems and working for the achievement of certain ideals is *one* way: there are others. To judge other, nonscientific, approaches by applying to them yardsticks proper to science is of

36. Ramm, *The Christian View of Science and Scripture*, p. 17.
37. Recall that in chapters 4 and 5 we described thinkers like Rescher and Laudan who understand the progress of science in antirealist terms. We will not rehearse that view here.

course question-begging. A successful argument for science being the paradigm of rationality must be based on the demonstration that the presuppositions of science are preferable to other presuppositions. That demonstration requires showing that science, relying on these presuppositions, is better at solving some problems and achieving some ideals than its competitors. But showing that cannot be the task of science. It is, in fact, one task of philosophy. Thus the enterprise of justifying the presuppositions of science by showing that with their help science is the best way of solving certain problems and achieving some ideals is a necessary precondition of the justification of science. Hence philosophy, and not science, is a stronger candidate for being the paradigm of rationality.[38]

Kekes's statement has been part of the theme of the first five chapters of this book, so little need be added here.

Second, we have also seen that several challenges have been raised against the progressive nature of science itself. Some, like Kuhn, have denied it altogether, opting for a replacement model and not a refinement one. Others allow for progress and refinement, but see a good deal less of it in the history of science than is commonly thought. For one thing, there are simply a number of examples of replacement in the history of science (oxygen theory replaced phlogiston theory). For another thing, various examples of vagueness have allowed a culture, already biased toward a realist view of scientific progress, to see more progress than is warranted.

Two examples of such vagueness are problems in how terms get their meaning and refer, and problems in specifying what counts as a later theory refining an earlier theory in the same domain and preserving that earlier theory within the later one, perhaps as a limiting case. Laudan and others have argued that when sufficient care is taken to clarify these issues, we find less progress than when these issues are allowed to be left unclarified. Left unclarified, transitional episodes tend to be given charitable readings along realist lines, due in part to the current prestige of science and scientific realism alike. For example, it is a substantive question whether Einstein's views of space, time, and matter refined or replaced Newton's theories. It is also a substantive question whether punctuated equilibrium theory refines or replaces classical or neo-Darwinianism.

Third, some antirealists grant that science has made progress in solving its problems, but they do not necessarily take that to imply that scientific imperialism is true. The progress of science, either in whole or, more likely, in part, can be understood on antirealist lines.

38. Kekes, *Nature of Philosophy* (Totowa, N.J.: Rowman and Littlefield: 1980), p. 158.

This opens the door to holding that a current successful scientific theory that conflicts with philosophy or theology should be understood in antirealist terms. At the very least, it indicates that progress in science does not entail that science is the most important voice in telling us what the world is truly like. Theology, philosophy, and other disciplines have a legitimate claim to be heard in the integrative venture.

Fourth, the nature and extent of progress depend on the level of generality of the system of concepts in question. Progress takes place within a paradigm that sets standards for the problems to be solved, the proper ways of solving them, and so on. The lower the level of generality exhibited by a paradigm, the more precisely it can be tested, falsified, verified, replaced, or refined. This is simply because, generally speaking, a low-level paradigm does not rest its epistemic support on a broad, complicated network of various kinds of evidence, epistemic values, and the like, at least in comparison with those groups of concepts that approach world view status or the level of broad research programs.

Two implications follow from this. First, there will be different degrees of progress in science at different levels of generality. Broad research programs will often coexist with rival research programs (e.g., vitalism versus physicalism, indeterminism versus determinism as a view of nature), and a given research program will tend to wane, rise, and cycle back and forth in ways not clearly indicative of progress.

Second, theology and especially philosophy tend to operate at higher levels of generality than does science.[39] So, in general, we should not expect theology or philosophy to progress as science does. Progress is not an appropriate standard for rational comparison between two theories or disciplines when these operate at different levels of generality.

For example, there are only a few broad theories of the nature of substance (Aristotle's theory, the Bundle theory, Locke's theory, and a few others), but one could take an endless number of progressive views about the detailed structure of, say, the substance carbon (as embodied in the various models of the atom). Greater progress in views of the substance carbon than in views of substance in general

---

39. This is not always the case. For example, philosophers sometimes focus on specific issues like an analysis of the moral duty of beneficence, and scientists sometimes focus on very general issues like the characterization of matter. Further, there are levels of generality within philosophy and theology, e.g., Protestant theology, Lutheran theology, covenant theology, a theology of the Abrahamic covenant, a theological treatment of a specific verse in Scripture. For a very useful discussion of science and philosophy with a treatment of levels of generality, see Schlesinger, *Metaphysics: Methods and Problems* (Totowa, N.J.: Barnes and Noble, 1983), pp. 9–26.

can be explained, in part, by the limited number of ultimate options available in philosophy and the fact that one of them, especially if the options are exhaustive, is true. Thus, progress is not necessarily a sign of epistemic virtue. It may be a sign that more options are available through which progress can be made at a certain level of generality than at another.

Fifth, if some philosophical or theological view is true, or some scientific one for that matter, we should not expect further progress in that area. Thus, progress can only be a sign of approximate truth at best, not of truth itself. Stanley Jaki has made this observation:

> . . . it will not be possible without risk to dismiss natural theology with the remark that, unlike science, natural theology shows no progress. Being in possession of a basically valid perspective, natural theology is under no obligation to make progress, just as science itself cannot outgrow its metaphysical matrix.[40]

The implication of this for comparing the rationality of science with that of philosophy or theology is actually the opposite of the claim we're responding to. The slow progress in philosophy and theology may indicate not that they are less rational than science—that is, that they have progressed less toward truth—but that they are more rational. Why? Because the slow progress could be an effect of their already having eliminated proportionately more false options in their spheres of study than science has eliminated in its. If this is true, it means that they have already come closer to a full, well-rounded, true world view than science has come.

An illustration might be helpful. Consider the process of writing, editing, and testing computer software. A programmer sets out with an initial goal in mind. He writes software to achieve that goal. Then he tests the software. His test reveals errors—what programmers call bugs—in the software, problems that, under certain conditions, cause malfunctions in the operation of the program. He edits the program to remove the bugs, then tests it again. This time he finds bugs again, but fewer. Each time he repeats the process, he finds fewer bugs, until at last there are none. What is relevant to our discussion is that early in the process progress is rapid—lots of bugs are found and eliminated. But the longer the process continues, and the closer it comes to resulting in a flawless program, the slower the progress becomes. Finally, when there are no more bugs, progress stops. Now, does the

---

40. Stanley Jaki, *The Road of Science and the Ways to God* (Chicago: University of Chicago Press, 1978), p. 327.

slower progress near the end of the process, or the complete lack of progress at its end, imply that the result—the flawless program—is less rational than the early program with all its flaws? Of course not. Neither, therefore, does the slower progress of philosophy or theology than of science necessarily indicate that those disciplines are less rational than science.

Indeed, theoretically at least (and in reality, according to traditional theism), there could be one mind that would make no progress at all: a mind that already knew the truth about absolutely everything past, present, and future. Whether such a mind exists is not the point here. The point is that the lack of progress in such a mind would evidence not its low level of rationality compared with minds that experienced rapid progress but the perfection of its rationality.

This does not mean that theologians or philosophers are dogmatic, narrow-minded bigots who disregard negative evidence. On the contrary, with some exceptions, theology and philosophy are characterized by rigorous argumentation and serious consideration of counterevidence. But if one thinks that one has finally reached a true conclusion, then even though it will always be possible to bring in new arguments and considerations, one can still trust the truth and rationality of the conclusion.

A distinction made by Roderick Chisholm is relevant here.[41] He distinguishes between two different senses of one's epistemic right to be sure about some belief. The first sense is a negative one and means that one terminates inquiry and will no longer be open to new evidence or arguments. The other sense implies that when one is sure of some conclusion, he can still be open to further investigation should new evidence arise, but in the meantime he has a right to claim to know the conclusion in question and to use it to ground inferences to other epistemic claims.

In sum, philosophy and theology may not progress because they may already have arrived rationally at some truth concerning the world. This means that a philosopher or theologian has the right to be sure about this conclusion, not in the sense of terminating inquiry or being closed to new arguments, but in the sense of requiring a good bit of evidence before abandoning the conclusion and not being able to use it to infer other conclusions.

Sixth, it is not true that philosophy and theology do not make progress. As I have already pointed out, philosophy and theology often work at a level of generality greater than that of science. But not

---

41. Roderick Chisholm, *Theory of Knowledge*, 2d ed. (Englewood Cliffs, N.J.: Prentice-Hall, 1977), pp. 116–18.

always. In fact, philosophy and theology operate with theories at various levels of generality. Furthermore, we have already seen that progress is possible with a paradigm that provides agreed-on problems and questions to be solved, standards for solution appraisal, and so on. Philosophy and theology do progress within paradigms. For example, contemporary utilitarians have advanced their views beyond those of Bentham and Mill, eighteenth- and nineteenth-century utilitarians. Contemporary dualism, like that embraced by followers of the phenomenology of Husserl, has advanced in clarity, problem-solving ability, and sophistication over the dualism of Descartes.

In both theology and philosophy, progress is often made with regard to original arguments, conceptual clarity, the refutation of some old arguments, and the clarification and refinement of others, especially when progress is measured within a particular philosophical paradigm, though entire disciplines of philosophy (e.g., epistemology) and theology (e.g., eschatology or ecclesiology) show progress. In general, the progress is less than that found in science, especially that in low-level scientific generalizations. But progress occurs in theology and philosophy nonetheless.

Finally, some of the progress of science can be explained by non-rational, sociological factors. A discipline will tend to progress if it is a cultural value such that resources go into that discipline and it attracts a number of bright minds. Science has progressed in the last few centuries, especially the last century, in part because of the cultural resources put into it. If those same resources had been put into philosophy or theology, then more progress might have obtained in those fields. The progress might not have been as dramatic or rapid as in science for reasons already discussed. But it might still have obtained. For example, since the formation of the Society of Christian Philosophers in 1978 there has been an extraordinary amount of progress in philosophy of religion, both in breadth and in sophistication.

In sum, the relative progress of science over theology and philosophy does not imply that the former is rational and the latter are not. Nor does it imply that in conflicts between science and philosophy or theology it is science that should get the best hearing. This leads us to our final consideration.

Creationism, Evolution, and Epistemic Values

It is often objected that creationism does not embody some specific epistemic value that a scientific theory ought to embody. We have already discussed predictability and empirical testability in this regard. But another epistemic value is often raised against creationism: fruitfulness in guiding future research. Douglas Futuyma complains that creationist literature fails to present "new information, or ideas for research that could resolve the issue in favor of creationism."[42]

We have already seen that, historically speaking, the creation/ evolution debate of the nineteenth century was not merely a debate about scientific facts. It was also a philosophical debate about what science should be like, a debate that can be understood in terms of a difference of opinion regarding the relative importance of epistemic values for scientific theories. In chapter 2 we listed several epistemic values—truth, simplicity, empirical accuracy, predictive success, internal consistency and clarity, adequacy in handling external conceptual problems, scope of reference, fruitfulness in guiding new research, and standards for what counts as science.

The current creation/evolution debate can be understood similarly. The current debate is certainly about scientific facts—fossils, the early earth's atmosphere, and so on. But it is also about epistemic values. Keeping this in mind can help us understand the differences between creationists and evolutionists in a way not possible if we focus merely on fossils and their kin. Each side claims to have certain epistemic values on its side. For example, both creationists and evolutionists claim to have truth (or approximate truth), both claim to have empirical accuracy, and both claim to have predictive success. I am not qualified to evaluate these claims; I leave that task to scientists. But each side at least claims equality or superiority regarding these epistemic values.

There are other epistemic values, however, that I do not think are shared equally by both theories.[43] On the one hand, evolutionists claim two key epistemic values that tip the scales in their favor: evolution accords (and creationism does not accord) with accepted standards for what science ought to be; evolution is fruitful in guiding new research.

We have already discussed the first issue, and I have argued that this debate is almost totally a second-order philosophical debate and not, strictly speaking, a scientific debate at all. I also have argued already against this modern version of the "positivist episteme," as Gillespie calls it.

---

42. Futuyma, *Science on Trial*, p. 178. See also Kitcher, *Abusing Science*, p. 126, 164.

43. I am omitting theoretical simplicity because it is not clear to me how this epistemic value is relevant to the creation/evolution debate and, in any case, I think that fruitfulness, philosophical and sociological standards for "good" science, and the role of conceptual problems are the key epistemic values at issue.

The second epistemic value, fruitfulness, seems to be more crucial.[44] Creationism is constantly challenged to make its claims more specific and precise so that detailed problems can be solved and future lines of investigation suggested.

Our earlier discussion of creationist predictions and retrodictions showed that creationism does suggest data for investigation. Nevertheless, evolutionists are basically right in this criticism. Creationism is less adequate than evolutionary theory regarding fruitfulness in guiding new research, and creationists' labor ought to be invested in this area.

On the other hand, creationists claim to be superior at handling internal and external conceptual problems. Consider the internal epistemic value of clarity. For example, Michael Denton argued recently that there are two sorts of evidence for evolution that do not depend on actual observations of the process: finding a sequence of forms that leads unambiguously from one form to another, and reconstructing a hypothetical pathway for such a transition that is clear, plausible, and convincing as an explanation for that transition.[45] Denton goes on to argue that in most cases the latter course must be followed, since no sequence of forms is available. He also argues that no clear pathway can be constructed in most of these cases (e.g., the transition from reptile scales to bird feathers).[46]

What Denton is doing in this and a host of other cases is faulting evolutionary theory with not embodying the epistemic value of inter-

---

44. Kitcher complains, "The theory [of creationism] has no infrastructure, no ways of articulating its vague central idea, so that specific features of living forms can receive detailed explanations." *Abusing Science*, p. 126. Crucial reading in this regard is Etienne Gilson's analysis of the species concept in philosophy and evolution. See *From Aristotle to Darwin and Back Again: A Journey in Final Causality, Species, and Evolution* (Notre Dame: University of Notre Dame Press, 1984). Gilson argues, among other things, that the notion of an essence and a final cause toward which something moves may not be fruitful to a pragmatically oriented descendant of Francis Bacon who seeks power and control over nature. But, he adds, that is no criticism against the truth of a notion. Truth may not be very practical or fruitful in guiding new ideas. I am inclined to agree. The notion of organisms having an essence, developing so as to realize a final cause, or the notion of God creating life, can be true without being particularly fruitful in suggesting lines of investigation for material or mechanical/efficient causes. But if a concept is to be scientific, remembering the relative unclarity of that term, somewhere or other it needs to move from the level of a general research tradition to a more specific, low level concept that suggests tests. Such a move does not necessarily transform the concept from philosophical (or theological) to scientific, but it does enhance the concept's scientific respectability.

45. See Denton, *Evolution*, pp. 55–56. As far as I know, Denton is not a creationist, but his ideas can obviously be used as part of a case for creationism.

46. See ibid., pp. 199–232 for other examples.

nal clarity. The same strategy is used when creationists argue against the probability of life arising spontaneously by natural processes, when they cite the incredible complexity of living organisms with designlike features (e.g., the detailed metamorphosis involved in the development of an adult butterfly), and when they appeal to the presence of information in DNA. The basic argument is that these provide internal conceptual problems for evolutionary theory. It is not clear, within evolutionary theory, how these problems can be solved, but creationists claim that the presence of a Mind can explain these phenomena.[47]

Creationists also raise external conceptual problems against evolutionary theory. Many of these are what Laudan calls world view difficulties.[48] Laudan has pointed out that there has never been a period in the history of thought when scientific theories exhausted the domain of rationality. World view difficulties are rationally justifiable problems for a scientific theory that derive from general world view considerations stemming largely from metaphysics, logic, ethics, and theology.

Examples of this kind of external conceptual problem are not hard to find. Creationists argue that if naturalistic evolution is all there is to the development of life, then it is hard to justify objective meaning in life, objective value and dignity for man, and so on. They also argue that it would be hard to see how the mind and senses could be trusted to give us truth about the world, a problem that plagued Darwin. I have discussed these problems elsewhere and do not wish to pursue them here.[49] But one point should be mentioned. Raising problems about the implications of evolutionary theory for mind, meaning, and value is not an ad hominem strategy; neither is it irrelevant for science. The idea that problems of this kind are irrelevant for science comes from a lack of appreciation of the role of external conceptual problems in the history of science and from the modern tendency toward analysis rather than synthesis and toward a fragmented, dichotomized world view where science is isolated from other disciplines.

If the mind is different from the brain, if thinking is a normative

47. For a comparison between scientific explanation, construed along the lines of a covering-law model, and personal explanation, see Richard Swinburne, "The Argument from Design," *Philosophy* 43 (July 1968): 199–212. I differ with Swinburne over the question of the scientific status of personal explanations. We saw in chapter 2 that a covering-law model of explanation is not the only one used in science. Specifically, scientists also explain by constructing models that postulate causes for effects. I see no reason why the postulated causes of such a scientific model cannot be personlike, as is often the case in psychology, sociology, and archaeology.

48. Laudan, *Progress and Its Problems* (Berkeley: University of California Press, 1977), pp. 61–64.

49. See Moreland, *Scaling the Secular City*, pp. 49–50, 77–132.

process involving full-blown freedom of the will, if there are objective values and objective meaning in life, and if good reasons can be given for believing in these notions, then they do constitute a set of external conceptual problems for naturalistic evolution. While I am not an advocate of theistic evolution—indeed, I think it is false—I do think it has the resources to respond to these external conceptual problems in a way not available to naturalistic evolutionary theory. So these external conceptual problems provide a set of difficulties for naturalistic evolutionary theory that can be solved by theistic evolution, special creationism, and progressive creationism.

There is, however, one further external conceptual problem that creationists raise against evolutionary theory. Evolutionary theory conflicts with rational theological claims. Despite all the emotional name calling and ad hominem attacks against creationists, and despite the unfounded protests that science and theology should not mix, it can be rational to raise theological problems against a scientific theory. To cite Laudan once more:

> Thus, contrary to common belief, it can be rational to raise philosophical and religious objections against a particular theory or research tradition, if the latter runs counter to a well-established part of our general *Weltbild*—even if that *Weltbild* is not "scientific" (in the usual sense of the word).[50]

Now it is certainly possible, indeed, actual, that some believers are irrational, Bible-thumping dogmatists who are fideists regarding their theological beliefs and uninformed about the intricacies of biblical hermeneutics and evolutionary theory alike. But ridicule and laughter are poor substitutions for argument. In this context, it is a boorish, irresponsible thing for a scholar to "argue" against creationism by calling creationists dogmatic fundamentalists, and so forth, a tactic engaged in by scholars like Ruse, Kitcher, Futuyma, and a host of others.

If understanding is to be reached, and if both sides are to learn from each other, such tactics must be abandoned. What is needed is a deeper understanding of the fact that creationists believe that external conceptual problems are relevant in scientific theory assessment and that they raise some serious external conceptual problems against evolutionary theory.

Suppose someone holds to these two propositions:

1. The Bible is the Word of God and it teaches the truth on matters of which it speaks.

50. Laudan, *Progress and Its Problems*, p. 124.

2. The Bible, properly interpreted, teaches (among other things) certain truths that run counter to evolutionary theory.

Suppose further that one has a list of rational arguments for these two propositions and does not merely embrace them by blind faith. In support of 1) he lists arguments from prophecy, history, archaeology, and other areas of science for the contention that it is rational to believe the Bible when it speaks on some matter, science included. In support of 2) he offers detailed arguments from hermeneutical theory, linguistics, comparative Near Eastern studies, and so forth.

In the case just cited, such an individual would have *reasons*, perhaps good reasons, for believing that the general theory of evolution will turn out to be false and that creationism will be vindicated. Such a person could be rational in believing such a position, even if certain anomalies present themselves to his or her theory from, say, genetics.

I do not wish to go further in this dialectic. I am aware that both of the propositions have been challenged. My point is that the evangelical community is not a group of irrational, snake-handling dogmatists with no knowledge of hermeneutics and biblical criticism who blindly embrace propositions 1 and 2. Some members of the evangelical community may behave that way, and some members of the evangelical community may not agree with 1 and 2 as I have stated them above. But a growing number of mature Christian thinkers, theologians and scientists alike, embrace 1 and 2 and assert good reasons for their beliefs. Furthermore, they offer their theological propositions as external conceptual problems that are appropriate considerations in evolutionary theory assessment.

In light of this, I suggest that more understanding would obtain if we read the creation/evolution debate not only as a difference regarding scientific facts, though it includes that, but also as a conflict over epistemic values. Specifically, creationists believe that the weight of conceptual problems, internal and external, is sufficient against evolutionary theory to warrant embracing creationism as a scientific research program. If I am right about this, then anyone who tries to interpret the creation/evolution debate merely along the lines of scientific facts will have a misguided, truncated view of the complexities of the intellectual issues involved. This situation is exacerbated by the general cultural demise of the humanities, the rise of scientism, and the naive compartmentalization that has resulted.

A better line of approach would be to continue to clarify the more empirical, scientific issues while simultaneously debating the general hermeneutical questions and apologetic issues involved in propositions 1 and 2. For example, evangelical Old Testament scholars need to work together with creation scientists to forge better and more defensible models. Furthermore, more work needs to be done in the area of external conceptual problems, especially as they have surfaced from theology. How have such problems figured in the history of science? When have theological problems helped science, when have they lost to science, when have they been complementary to science, and when have they defeated scientific ideas that were their rivals? These questions are part of the debate about creation/evolution seen in its proper intellectual perspective.

Summary

In this chapter we have investigated the scientific status of creation science. We began with a brief sketch of the historical background to the current creation/evolution debate and moved on to consider a working characterization of creation science. Next, objections against the scientific status of creation science were considered. This led to an investigation of the claim that science is rationally superior to theology or philosophy because it is more progressive. Finally, we investigated the creation/evolution debate from the perspective of epistemic values.

Concluding Unscientific Postscript

Thirty-five years ago, theologian and philosopher of science Bernard Ramm issued this challenge regarding Christian thinking about science:

> Orthodoxy did not have a well-developed philosophy of science or philosophy of biology. The *big* problems of science and biology must be argued in terms of a broad philosophy of science. The evangelical always fought the battle on too narrow a strip. He argued over the authenticity of this or that bone; this or that phenomenon in a plant or animal; this or that detail in geology. The empirical data is [sic] just there and the scientists can run the evangelical to death in constantly turning up new material. The evangelicals by fighting on such a narrow strip simply could not compete with the scientists who were spending their lifetime routing out matters of fact.[1]

Things have not changed appreciably since Ramm wrote those words. As I pointed out in the introduction, a number of thinkers like R. C. Sproul and Charles Malik have called this the most anti-intellectual period in the church. Even if that claim is wrong, there can be no question—granted the books we buy, the level of thinking we exemplify as a total community, the impact we have had on the thinking of the world—that Christians have valued personal piety and practical application *at the expense of embracing serious study as a Christian cultural value.*

This anti-intellectualism has led to fideism, a secular/sacred dichotomy in our view of Christian life and witness, inordinate individualism, and resort to simplistic slogans in the place of serious answers to problems, to name a few sad results. These features of the Christian community have come at a most inopportune time. The rapid rise of

1. Bernard Ramm, *The Christian View of Science and Scripture* (Grand Rapids: Eerdmans, 1954), p. 18.

science as a set of explanatory paradigms and as a prestige-bearing sociological phenomenon has met with little understanding or inter-action in the Christian community at large. There are important and noticeable exceptions to this rule, but those exceptions do not receive the support or have the influence they deserve in the Christian community.

But things appear to be changing. In recent years there has been a noticeable increase in the number of serious Christian thinkers who are professional philosophers, theologians, biblical scholars, and scientists. This increase has been coupled with a number of discoveries in science that are favorable to Christianity (e.g., the Big Bang theory and information in DNA). These are exciting days and no time for retreat. Instead, we need to embrace more fully as a broad, Christian cultural value the importance of study, scholarship, and the mind for the health and mission of the Christian church. This is especially true in the area of the interaction of science and theology.

As we have seen, the integration of science and theology must draw from several areas of study. Thus, it must be a community affair. Biblical scholars, theologians, scientists, philosophers, and historians and sociologists of science are all important in the community task of forging better models of science/theology integration that are faithful to the propositional revelation of God in Scripture and responsible to the best insights from the disciplines.

This book has been an attempt to show that philosophy of science has much to contribute to the interface between science and theology. In chapter 1, we saw that there is no definition of science, no set of necessary and sufficient conditions that can be given for drawing a line of demarcation between science and nonscience. Protests to the contrary are simply uninformed. Philosophy and theology were seen to interface with science in a number of ways.

In chapter 2, we considered the objection that science differs from nonscience in that the former possesses a method the latter lacks, namely, *the* scientific method. We investigated different views of the scientific method, inductivism, and an eclectic model of scientific methodology, and we investigated seven different areas of debate involved in fleshing out the details of an eclectic model. This study led to the conclusion that there is no such thing as the scientific method. Rather, there is a family of scientific methodologies. Furthermore, we saw that fields outside of science use, in one degree or another, different aspects of scientific methodology. Thus, the so-called scientific method fails to be a criterion for drawing a line between science and nonscience.

Chapters 1 and 2 opened the door for philosophical and theological considerations to enter into science. Chapter 3 considered a protest from scientism to the effect that, in contrast to philosophy, theology, and other disciplines, science is the only rational, truth-securing approach to the world. Various kinds of limits to science were investigated and the conclusion was reached that the conversation between science and philosophy/theology is a dialogue, not a monologue.

In chapters 4 and 5, we turned our attention to the debate between scientific realism and antirealism. Our purpose was to promote an eclectic approach to the integration of science and theology that adopts a realist or an antirealist view of science on a case-by-case basis and uses the insights of this debate as part of the integrative task. Chapter 4 surveyed several objections to scientific realism. Chapter 5 overviewed different versions of antirealism and offered an admittedly brief sketch of an eclectic model of science itself.

Finally, in chapter 6 we investigated the second-order philosphical question of whether creation science should be classified as science instead of religion. We placed the present discussion in its historical context, looked at a working characterization of creation science, and responded to charges raised against the scientific status of creation science. In the process, we looked at how the prestige of science—derived from its progressive nature—enters into debates between science and theology. We also looked at a way of viewing that debate between creation and evolution as a conflict between different epistemic values held to be desirable for a scientific theory.

A good bit of work still needs to be done, and the Christian community needs to develop its philosophy of science, as Ramm said years ago. In my opinion, we need to do three things.

First, we need to change our view in the local church about what fellowship means. Too often it means getting together to chat and eat food. As important as this might be, the New Testament understanding of fellowship involves gathering together for mutual encouragement and the practice of spiritual gifts *with a view toward equipping the body to be more successful in promoting the progress of the gospel.* If I am right, then I suspect that more of our fellowship should include honest, thought-provoking investigations of our own theological beliefs as they interact with the world we are trying to reach. Specifically, science and the philosophy of science should be more a part of local church fellowship than they currently are, if we really are to meet together to be more effective in reaching the world and if scientism is a major factor constituting the world we are to reach.

Second, Christian parents need to recapture the theological notion of a vocation—a life message, a calling by God to use spiritual gifts, natural talents, and circumstances of life for the progress of the gospel to the glory of God. Such a notion of vocation goes well beyond a job as a means of securing resources to meet one's own needs and the needs of those who proclaim the gospel. If Christian parents recapture the notion of vocation, then they will encourage their children to go to college not primarily to secure the skills to get a job but, more importantly, to find a way of expressing their vocation better through learning to think more critically about some area of university study. If this happens, I suspect more and more Christians will go into the humanities, philosophy and theology included. This is not to eschew other fields, but there is a great need for believers to master the humanities to serve Christ. Philosophy and history of science are two fields of crucial importance for responsible Christians to take up.

Third, we need to give more of our money to support the efforts of Christian thinkers who are working on the integration of science and theology and are engaged in apologetics. This point could, of course, be extended to include Christian education in general—especially colleges, seminaries, and universities. But in any case, if scientism is a major factor in the world today, and if we are to reach that world, then more resources need to go to those attempting to accomplish that task.

These are only suggestions, and readers may disagree with them or have better ones. But if the arguments in this book are reasonable, one thing seems clear. Science and theology can and should interact in various ways, as they have done throughout history. The current compartmentalization of science and religion is wrong, as is the idea that science is an isolated set of disciplines. Further, scientism is a false, dehumanizing view of the world. Christians can and must think responsibly about how to relate science and theology. When they do, philosophy will be an essential part of that enterprise.

Select Bibliography

A. General Books in the Philosophy of Science

Achinstein, Peter. *Concepts of Science: A Philosophical Analysis.* Baltimore: Johns Hopkins, 1968.

Ackermann, Robert John. *Data, Instruments, and Theory: A Dialectical Approach to Understanding Science.* Princeton, N.J.: Princeton University Press, 1985.

Bunge, Mario. *Method, Model, and Matter.* Dordrecht, Holland: D. Reidel, 1973.

Danto, Arthur Coleman, and Sidney Morgenbesser, eds. *Philosophy of Science.* New York: Meridian, 1960.

Feigl, Herbert, and May Brodbeck. *Readings in the Philosophy of Science.* New York: Appleton-Century-Crofts, 1953.

Harre, Rom. *The Philosophies of Science: An Introductory Survey.* Oxford: Oxford University Press, 1972.

Hempel, Carl G. *Aspects of Scientific Explanation.* New York: Macmillan, 1965.

_____. *Philosophy of Natural Science.* Englewood Cliffs, N.J.: Prentice-Hall, 1966.

Kourany, Janet A. *Scientific Knowledge: Basic Issues in the Philosophy of Science.* Belmont, Calif.: Wadsworth, 1987.

Lambert, Karel, and Gordon G. Brittan, Jr., eds. *An Introduction to the Philosophy of Science.* 3d ed. Atascadero, Calif.: Ridgeview, 1987.

Mannoia, V. James, Jr. *What is Science? An Introduction to the Structure and Methodology of Science.* Lanham, Md.: University Press of America, 1980.

Nagel, Ernest. *The Structure of Science.* Indianapolis: Hackett, 1979.

Popper, Karl. *Conjectures and Refutations: The Growth of Scientific Knowledge.* New York: Harper and Row, 1963.

Ratzsch, Del. *Philosophy of Science.* Edited by C. Stephen Evans. Downers Grove: Inter-Varsity, 1986.

Salmon, Wesley. *The Foundations of Scientific Inference.* Pittsburgh: University of Pittsburgh Press, 1967.

Shapere, Dudley, ed. *Philosophical Problems of Natural Science*. New York: Macmillan, 1965.

Suppe, Frederick, ed. *The Structure of Scientific Theories*. 2d ed. Urbana, Ill.: University of Illinois Press, 1977.

For overviews of philosophy of science, Ratzsch, Mannoia, and Hempel (Prentice-Hall) are good places to start. A bit more difficult, but more thorough, overviews are Harre, Kourany, and Lambert and Brittan.

B. Theological Method and Scientific Method

Macintosh, Douglas Clyde. *Theology As an Empirical Science*. New York: Macmillan, 1919.

Montgomery, John Warwick. "The Theologian's Craft." In *The Suicide of Christian Theology*, 267–313. Minneapolis: Bethany, 1970.

Pannenberg, Wolfhart. *Theology and the Philosophy of Science*. Translated by Francis McDonagh. Philadelphia: Westminster, 1973.

Schlesinger, George. *Religion and Scientific Method*. Dordrecht, Holland: D. Reidel, 1977.

Schoen, Edward L. *Religious Explanations: A Model from the Sciences*. Durham, N.C.: Duke University Press, 1985.

Sherry, Patrick. *Spirit, Saints, and Immortality*. Albany: State University of New York Press, 1984.

Montgomery is the best place to start. Schlesinger and Schoen are more difficult but would be the next ones to read after Montgomery.

C. General Treatments of Theology and Science

Barbour, Ian G. *Issues in Science and Religion*. New York: Harper and Row, 1966.

Bube, Richard. *The Human Quest: A New Look at Science and the Christian Faith*. Waco: Word, 1971.

Bube, Richard, ed. *The Encounter Between Christianity and Science*. Grand Rapids: Eerdmans, 1968.

Clark, Gordon. *The Philosophy of Science and Belief in God*. Nutley, N.J.: Craig, 1964.

Dooyeweerd, Herman. *The Secularization of Science*. Memphis, Tenn.: Christian Studies Center, 1954.

Henderson, Charles P., Jr. *God and Science: The Death and Rebirth of Theism*. Atlanta: John Knox, 1986.

Jaki, Stanley. *The Road of Science and the Ways to God.* Chicago: University of Chicago Press, 1978.

Mascall, Eric L. *Christian Theology and Natural Science: Some Questions on Their Relations.* London: Longmans, Green, 1956.

Morris, Henry. *The Biblical Basis of Modern Science.* Grand Rapids: Baker, 1984.

Peacocke, Arthur R. *Creation and the World of Science: The Bampton Lecturers.* Oxford: Oxford University Press, 1979.

Peacocke, Arthur R., ed. *The Sciences and Theology in the Twentieth Century.* Notre Dame: University of Notre Dame Press, 1986.

Polkinghorne, John C. *One World: The Interaction of Science and Theology.* Princeton, N.J.: Princeton University Press, 1987.

―――――. *The Way the World Is.* Grand Rapids: Eerdmans, 1984.

Ramm, Bernard. *A Christian View of Science and Scripture.* Grand Rapids: Eerdmans, 1954.

Rolston, Holmes, III. *Science and Religion: A Critical Survey.* Philadelphia: Temple University Press, 1986.

Smith, Wolfgang. *Cosmos and Transcendence: Breaking Through the Barriers of Scientific Belief.* Peru, Ill.: Sherwood Sugden, 1984.

Van Till, Howard J., Davis A. Young, and Clarence Menninga. *Science Held Hostage: What's Wrong with Creation Science and Evolutionism.* Downers Grove: Inter-Varsity, 1988.

These works are written from a variety of viewpoints: all have valuable insights.

D. History of Science: General

Boas, Marie. *The Scientific Renaissance: Fourteen Fifty to Sixteen Thirty.* New York: Harper and Row, 1962.

Bynum, W. F., E. J. Browne, and Roy Porter, eds. *Dictionary of the History of Science.* Princeton, N.J.: Princeton University Press, 1984.

Cohen, I. Bernard. *Revolution in Science.* Cambridge: Harvard University Press, 1985.

Dampier, William Cecil. *A History of Science.* New York: Macmillan, 1943.

Hall, A. R. *The Scientific Revolution: 1500–1800.* Boston: Beacon, 1954.

Harre, Rom. *Great Scientific Experiments: Twenty Experiments that Changed Our View of the World.* Oxford: Oxford University Press, 1983.

Kearney, Hugh. *Science and Change: 1500–1700.* New York: McGraw-Hill, 1971.

Mason, Stephen F. *A History of the Sciences.* Original title: *Main Currents of Scientific Thought.* New York: Collier, 1956.

Westfall, Richard S. *The Construction of Modern Science*. Cambridge: Cambridge University Press, 1978.

Mason surveys the entire history of science from ancient Babylon to the present, Harre is an interesting and readable discussion of twenty important scientific experiments, and Cohen is a massive volume that describes the role and nature of revolutions in the history of science. The dictionary by Bynum, et al., is a valuable tool.

E. History of Physical Science

Cohen, I. Bernard. *The Birth of A New Physics*. Rev. and expanded ed. New York: W. W. Norton, 1985.

Crosland, Maurice P., ed. *The Science of Matter: A Historical Survey*. Baltimore: Penguin, 1971.

Grant, Edward. *Physical Science in the Middle Ages*. Cambridge: Cambridge University Press, 1977.

Harman, P. M. *Energy, Force, and Matter: The Conceptual Development of Nineteenth-Century Physics*. Cambridge: Cambridge University Press, 1982.

Weaver, Jefferson Hane, ed. *The World of Physics: A Small Library of the Literature of Physics From Antiquity to the Present*. 3 vols. New York: Simon and Schuster, 1987.

F. History of Biology

Coleman, William R. *Biology in the Nineteenth Century: Problems of Form, Function, and Transformation*. Cambridge: Cambridge University Press, 1977.

Lanham, Url. *Origins of Modern Biology*. New York: Columbia University Press, 1968.

G. History of Evolution and Creation

Gale, Barry G. *Evolution Without Evidence: Charles Darwin and The Origin of Species*. Albuquerque: University of New Mexico Press, 1982.

Gillespie, Neal C. *Charles Darwin and The Problems of Creation*. Chicago: University of Chicago Press, 1979.

Greene, John C. *Science, Ideology, and World View: Essays in the History of Evolutionary Ideas*. Berkeley: University of California Press, 1981.

Morris, Henry. *History of Modern Creationism*. San Diego: Master, 1984.

Gillespie is the best one to get for an understanding of what was happening during the time of Darwin.

H. History of the Relationship Between Science and Theology

Hooykaas, Reijer. *Religion and the Rise of Modern Science*. Rev. ed. Edinburgh: Scottish Academic Press, 1973.

Lindberg, David C., and Ronald L. Numbers, eds. *God and Nature: Historical Essays on the Encounter Between Christianity and Science*. Berkeley: University of California Press, 1986.

Russell, Colin A. *Cross-Currents: Interactions Between Science and Faith*. Grand Rapids: Eerdmans, 1985.

All three are very valuable. Russell is the best place to start; Numbers and Lindberg is a thorough and recent work.

I. History of the Philosophy of Science

Losee, John. *An Historical Introduction to the Philosophy of Science*. 2d ed. Oxford: Oxford University Press, 1980.

J. Works Relating to Limits of Science

Bealer, George. "The Philosophical Limits of Scientific Essentialism." In *Philosophical Perspectives, Vol. 1: Metaphysics, 1987*, edited by James E. Tomberlin, 289–365. Atascadero, Calif.: Ridgeview, 1987.

Broad, William, and Nicholas Wade. *Betrayers of the Truth: Fraud and Deceit in the Halls of Science*. New York: Simon and Schuster, 1982.

Burtt, Edwin Arthur. *The Metaphysical Foundations of Modern Physical Science: A Historical and Critical Essay*. Rev. ed. Garden City, N.Y.: Doubleday, 1952.

Engelhardt, H. Tristram, Jr., and Daniel Callahan. *Morals, Science, and Sociality*. Hastings-on-Hudson, N.Y.: The Hastings Center, 1978.

Gurwitsch, Aron. *Phenomenology and the Theory of Science*. Edited by Lester Embree. Evanston, Ill.: Northwestern University Press, 1974.

Husserl, Edmund. *The Crisis of European Sciences and Transcendental Phenomenology: An Introduction to Phenomenological Philosophy*. Translated by David Carr. Evanston, Ill.: Northwestern University Press, 1970.

Kekes, John. *The Nature of Philosophy.* Totowa, N.J.: Rowman and Littlefield, 1980.

Losee, John. *Philosophy of Science and Historical Enquiry.* Oxford: Oxford University Press, 1987.

Medawar, Peter B. *The Limits of Science.* New York: Harper and Row, 1984.

Rescher, Nicholas. *The Limits of Science.* Berkeley: University of California Press, 1984.

Schlesinger, George N. *The Intelligibility of Nature.* Aberdeen, Scotland: Aberdeen University Press, 1985.

————. *Metaphysics: Methods and Problems.* Totowa, N.J.: Barnes and Noble, 1983.

Wechsler, Judith, ed. *On Aesthetics in Science.* Cambridge, Mass.: MIT Press, 1978.

Kekes is the best place to start. Rescher is more technical and Schlesinger and Bealer are more technical still, but all are valuable. Losee attempts to relate the different ways the history of science interfaces with the philosophy of science.

K. Realism and Antirealism

Churchland, Paul M., and Clifford A. Hooker, eds. *Images of Science: Essays on Realism and Empiricism, with a Reply from Bas C. van Fraassen.* Chicago: University of Chicago Press, 1985.

French, Peter A., Theodore E. Uehling, Jr., and Howard K. Wettstein, eds. *Realism and Antirealism.* Minneapolis: University of Minnesota Press, 1988.

Hanson, N. R. *Patterns of Discovery: An Inquiry into the Conceptual Foundations of Science.* Cambridge: Cambridge University Press, 1958.

Harre, Rom. *Varieties of Realism: A Rationale for the Natural Sciences.* Oxford: Basil Blackwell, 1986.

Hubner, Kurt. *Critique of Scientific Reason.* Translated by Paul R. Dixon and Hollis M. Dixon. Chicago: University of Chicago Press, 1983.

Kuhn, Thomas. *The Structure of Scientific Revolutions.* 2d ed., enlarged. Chicago: University of Chicago Press, 1970.

Lakatos, Imre, and Alan Musgrave, eds. *Criticism and the Growth of Knowledge.* Cambridge: Cambridge University Press, 1970.

Laudan, Larry. *Progress and Its Problems: Towards a Theory of Scientific Growth.* Berkeley: University of California Press, 1977.

————. *Science and Values: An Essay on the Aims of Science and Their Role in Scientific Debate.* Berkeley: University of California Press, 1984.

————. "Explaining the Success of Science: Beyond Epistemic Realism and Relativism." In *Science and Reality,* edited by James T. Cushing, C. F.

Delaney, and Gary M. Gutting, 83–105. Notre Dame: University of Notre Dame Press, 1984.

Leplin, Jarrett, ed. *Scientific Realism.* Berkeley: University of California Press, 1984.

Newton-Smith, W. H. *The Rationality of Science.* Boston: Routledge and Kegan Paul, 1981.

Papineau, David. *Theory and Meaning.* Oxford: Clarendon, 1979.

Putnam, Hilary. *Philosophical Papers, Volume 3: Reason, Truth, and History.* Cambridge: Cambridge University Press, 1981.

————. *The Many Faces of Realism.* La Salle, Ill.: Open Court, 1987.

Reichenbach, Hans. *The Rise of Scientific Philosophy.* Berkeley: University of California Press, 1951.

Rorty, Richard. *Philosophy and the Mirror of Nature.* Princeton, N.J.: Princeton University Press, 1979.

Smart, J. J. C. *Philosophy and Scientific Realism.* London: Routledge and Kegan Paul, 1963.

van Fraassen, Bas C. *The Scientific Image.* Oxford: Oxford University Press, 1980.

All the works in this area are tough going, but Newton-Smith is probably the best introduction and overview. Kuhn is a benchmark and the works by Laudan are clear and easy to read. Leplin is more technical but covers several areas of current debate.

L. The Scientific Status of Creation Science

Futuyma, Douglas. *Science on Trial.* New York: Pantheon, 1982.

Geisler, Norman L. *The Creator in the Courtroom.* Milford, Mich.: Mott Media, 1982.

Geisler, Norman L., and J. Kerby Anderson. *Origin Science.* Grand Rapids: Baker, 1987.

Kitcher, Philip. *Abusing Science: The Case Against Creationism.* Cambridge, Mass.: MIT Press, 1984.

Newell, Norman D. *Creation and Evolution: Myth or Reality?* New York: Praeger, 1985.

Quinn, Philip. "The Philosopher of Science as Expert Witness." In *Science and Reality,* edited by James T. Cushing, C. F. Delaney, and Gary M. Gutting, 32–53. Notre Dame: University of Notre Dame Press, 1984.

Ruse, Michael. *Darwinism Defended.* Reading, Mass.: Addison-Wesley, 1982.

For criticisms of the scientific status of creation science, see Kitcher, Ruse, Futuyma, and Newell. For defenses of the scientific status of creation science see Geisler, Quinn, and the articles by Laudan cited in chapter 6 of the present work.

M. Scientific Criticisms of Evolution

Byrd, W. R. *Origin of the Species Revisited.* 2 vols. New York: Philosophical Library, 1987.

Denton, Michael. *Evolution: A Theory in Crisis.* London: Burnett Books, 1985.

Gentry, Robert V. *Creation's Tiny Mystery.* Knoxville, Tenn.: Earth Science Associates, 1986.

Gish, Duane. *Evolution: The Challenge of the Fossil Record.* El Cajon, Calif.: Master, 1985.

Hitching, Francis. *The Neck of the Giraffe: Darwin, Evolution, and the New Biology.* New York: Mentor, 1982.

Lester, Lane P., and Raymond G. Bohlin. *The Natural Limits to Biological Change.* Grand Rapids: Zondervan, 1984.

Morris, Henry M. *Scientific Creationism.* El Cajon, Calif.: Master, 1985.

————. *Creation and the Modern Christian.* El Cajon, Calif.: Master, 1985.

Morris, Henry, and Gary Parker. *What Is Creation Science?* San Diego: Creation Life Publishers, 1982.

Pitman, Michael. *Adam and Evolution: A Scientific Critique of Neo-Darwinism.* London: Rider and Company, 1984.

Pun, Pattie P. T. *Evolution: Nature and Scripture in Conflict?* Grand Rapids: Zondervan, 1982.

Shapiro, Robert. *Origins: A Skeptic's Guide to the Genesis of Life on Earth.* New York: Summit, 1986.

Taylor, Gordon Rattray. *The Great Evolution Mystery.* New York: Harper and Row, 1983.

Thaxton, Charles B., Walter L. Bradley, and Roger L. Olsen. *The Mystery of Life's Origin: Reassessing Current Theories.* New York: Philosophical Library, 1984.

Thurman, L. Duane. *How to Think about Evolution.* Rev. ed. Downers Grove: Inter-Varsity, 1978.

Wiester, John. *The Genesis Connection.* Nashville: Thomas Nelson, 1983.

Wilder-Smith, A. E. *Man's Origin, Man's Destiny.* Wheaton: Harold Shaw, 1968.

Thurman or Morris (*Scientific Creationism*) is a good place to start. Denton, Gish, and Thaxton, Bradley, and Olsen cover most of the bases in a way that is technically respectable. I suggest these three as the best of the lot.

N. Philosophical and Theological Interactions with Physical Science

Cartwright, Nancy. *How the Laws of Physics Lie.* Oxford: Oxford University Press, 1983.

Davies, P. C. W. *The Accidental Universe*. Cambridge: Cambridge University Press, 1982.

_____ . *God and the New Physics*. New York: Simon and Schuster, 1983.

Friedman, Michael. *Foundations of Space-Time Theories: Relativistic Physics and Philosophy of Science*. Princeton, N.J.: Princeton University Press, 1983.

Heisenberg, Werner. *Physics and Philosophy: The Revolution in Modern Science*. New York: Harper and Row, 1958.

Herbert, Nich. *Quantum Reality*. Garden City, N.Y.: Doubleday, 1985.

McMullin, Ernan, ed. *The Concept of Matter in Greek and Medieval Philosophy*. Notre Dame: University of Notre Dame Press, 1963.

_____ , ed. *The Concept of Matter in Modern Philosopy*. Notre Dame: University of Notre Dame Press, 1963.

Morris, Richard. *The Nature of Reality*. New York: McGraw-Hill, 1987.

Stafleu, Marinus Dirk. *Time and Again*. Toronto: Wedge, 1980.

Stebbing, L. Susan. *Philosophy and the Physicists*. New York: Dover, 1958.

Swinburne, Richard, ed. *Space, Time, and Causality*. Dordrecht, Holland: D. Reidel, 1983.

All the works listed here are fairly technical. For introductions to issues in this area, see Barbour or Rolston in section C.

O. Philosophical and Theological Interactions with Biological Science: General

Ayala, Francisco, and Theodosius Dobzhansky, eds. *Studies in the Philosophy of Biology: Reduction and Related Problems*. Berkeley: University of California Press, 1974.

Greene, Marjorie, ed. *Approaches to a Philosophical Biology*. New York: Basic, 1965.

Hull, David. *Philosophy of Biological Science*. Englewood Cliffs, N.J.: Prentice-Hall, 1974.

Medawar, Peter B., and J. S. Medawar. *Aristotle to Zoos: A Philosophical Dictionary of Biology*. Cambridge: Harvard University Press, 1983.

Peacocke, Arthur R. *God and the New Biology*. San Francisco: Harper and Row, 1986.

Smith, Vincent Edward, ed. *Philosophical Problems in Biology*. New York: St. John's University Press, 1966.

Hull is a good place to start and Peacocke is a helpful introduction as well.

P. Philosophical and Theological Interactions with Biological Science: Functional Explanation, Design, and Teleology

Achinstein, Peter. *The Nature of Explanation*, chapter 8. Oxford: Oxford University Press, 1983.

Barrow, John D., and Frank J. Tipler. *The Anthropic Cosmological Principle*. Oxford: Clarendon, 1986.

Dawkins, Richard. *The Blind Watchmaker: Why the Evidence of Evolution Reveals a Universe Without Design*. New York: W. W. Norton, 1986.

Gilson, Etienne. *From Aristotle to Darwin and Back Again: A Journey in Final Causality, Species, and Evolution*. Notre Dame: University of Notre Dame Press, 1984.

Horigan, James E. *Chance or Design?* New York: Philosophical Library, 1979.

Moreland, J. P. *Scaling the Secular City: A Defense of Christianity*, chapter 2. Grand Rapids: Baker, 1987.

Thorpe, William Homan. *Purpose in a World of Chance: A Biologist's View*. Oxford: Oxford University Press, 1978.

Wright, Larry. *Teleological Explanations: An Etiological Analysis of Goals and Functions*. Berkeley: University of California Press, 1976.

See Achinstein for an introduction to functional explanation and Moreland for an introduction to the design argument. Gilson is must reading for anyone wanting to compare efficient and final causality.

Q. Philosophical and Theological Interactions with Biological Science: Evolutionary Theory

Blocher, Henri. *In the Beginning: The Opening Chapters of Genesis*. Downers Grove: Inter-Varsity, 1984.

Delbruck, Max. *Mind From Matter: An Essay on Evolutionary Epistomology*. Edited by Gunther Stent and David Presti. Oxford: Basil Blackwell, 1986.

Jaki, Stanley L. *Angels, Apes, and Men*. La Salle, Ill.: Sherwood Sugden, 1983.

Jensen, Uffe J., and Rom Harre, eds. *The Philosophy of Evolution*. New York: St. Martin's, 1981.

McMullin, Ernan, ed. *Evolution and Creation*. Notre Dame: University of Notre Dame Press, 1985.

Morris, Henry. *King of Creation*. San Diego: Creation Life Publishers, 1980.

Murphy, Jeffrie G. *Evolution, Morality, and the Meaning of Life*. Totowa, N.J.: Rowman and Littlefield, 1982.

Newman, Robert C., and Herman J. Eckelmann, Jr. *Genesis One and the Origin of the Earth*. Downers Grove: Inter-Varsity, 1977.

Schuster, George N., and Ralph E. Thorson. *Evolution in Perspective*. Notre Dame: University of Notre Dame Press, 1970.

Sober, Elliot. *The Nature of Selection: Evolutionary Theory in Philosophical Focus*. Cambridge, Mass.: MIT Press, 1984.

Sober, Elliot, ed. *Conceptual Issues in Evolutionary Biology: An Anthology*. Cambridge, Mass.: MIT Press, 1984.

Youngblood, Ronald, ed. *The Genesis Debate: Persistent Questions About Creation and the Flood*. Nashville: Thomas Nelson, 1986.

Youngblood contains a series of debates about several exegetical issues in Genesis and is quite helpful. The volumes by Sober are technical. See also Gilson in section P.

R. General Issues in Reductionism

Connell, Richard J. *Substance and Modern Science*. Houston: Center for Thomistic Studies, and Notre Dame: University of Notre Dame Press, 1988.

MacKay, Donald M. *The Clock Work Image: A Christian Perspective on Science*. Downers Grove: Inter-Varsity, 1974.

————. *Human Science and Human Dignity*. Downers Grove: Inter-Varsity, 1979.

Nagel, Ernest. *The Structure of Science*, chapters 11 and 12. Indianapolis: Hackett, 1979.

Spector, Marshall. *Concepts of Reduction in Physical Science*. Philadelphia: Temple University Press, 1978.

Van Leeuwen, Mary Stewart. *The Person in Psychology: A Contemporary Appraisal*. Edited by Carl F. H. Henry. Grand Rapids: Eerdmans, 1985.

The works listed rely exclusively on a complementarity approach to reductionism with the exception of Van Leeuwen and Connell. Connell is best because he correctly addresses questions of reductionism within a broad context where the philosophical notions of substance and property are addressed.

S. Works Relating to the Mind/Body Problem

Baker, Lynne Rudder. *Saving Belief: A Critique of Physicalism*. Princeton, N.J.: Princeton University Press, 1987.

Block, Ned, ed. *Readings in Philosophy of Psychology: Volume I*. Cambridge: Harvard University Press, 1983.

Boyle, Joseph M. Jr., Germain Grisez, and Olaf Tollefsen. *Free Choice: A Self-Referential Argument.* Notre Dame: University of Notre Dame Press, 1976.

Broad, C. D. *The Mind and Its Place in Nature.* London: Kegan Paul, 1925.

Campbell, Charles Arthur. *In Defense of Free Will: With Other Philosophical Essays.* London: George Allen and Unwin, 1967.

————. *On Selfhood and Godhood.* London: George Allen and Unwin, 1957.

Campbell, Keith. *Body and Mind.* Notre Dame: University of Notre Dame Press, 1970.

Chisholm, Roderick. *The First Person: An Essay on Reference and Intentionality.* Minneapolis: University of Minnesota Press, 1982.

Churchland, Paul M. *Matter and Consciousness.* Cambridge, Mass.: MIT Press, 1984.

————. *Scientific Realism and the Plasticity of Mind.* Cambridge: Cambridge University Press, 1979.

Collins, Arthur W. *The Nature of Mental Things.* Notre Dame: University of Notre Dame Press, 1987.

Dennet, Daniel C. *Elbow Room: The Varieties of Free Will Worth Wanting.* Cambridge, Mass.: MIT Press, 1984.

Ducasse, C. J. *Nature, Mind, and Death.* La Salle, Ill.: Open Court, 1951.

French, Peter A., Theodore E. Uehling Jr., and Howard K. Wettstein. *Midwest Studies in Philosophy X: Studies in the Philosophy of Mind.* Minneapolis: University of Minnesota Press, 1986.

Lewis, Hywel D. *The Elusive Mind.* London: George Allen and Unwin, 1969.

————. *The Elusive Self.* Philadelphia: Westminster, 1982.

————. *The Self and Immortality.* New York: Seabury, 1973.

Lovejoy, Arthur D. *The Revolt Against Dualism.* La Salle, Ill.: Open Court, 1955.

Lucas, John R. *The Freedom of the Will.* Oxford: Oxford University Press, 1970.

Madell, Geoffrey. *The Identity of the Self.* Edinburgh: Edinburgh University Press, 1981.

Moreland, J. P. *Scaling the Secular City: A Defense of Christianity,* chapter 3. Grand Rapids: Baker, 1987.

Pols, Edward. *The Acts of Our Being: A Reflection on Agency and Responsibility.* Amherst, Mass.: University of Massachusetts Press, 1982.

Popper, Karl, and John Eccles. *The Self and Its Brain: An Argument for Interactionism.* London: Routledge and Kegan Paul, 1977.

Robinson, Howard. *Matter and Sense: A Critique of Contemporary Materialism.* Cambridge: Cambridge University Press, 1982.

Searle, John. *Minds, Brains, and Science.* Cambridge: Harvard University Press, 1985.

Shaffer, Jerome. *Philosophy of Mind.* Englewood Cliffs, N.J.: Prentice-Hall, 1968.

Shoemaker, Sydney, and Richard Swinburne. *Personal Identity.* Oxford: Basil Blackwell, 1984.

Smith, Peter, and O. R. Jones. *The Philosophy of Mind: An Introduction.* Cambridge: Cambridge University Press, 1986.

Swinburne, Richard. *The Evolution of the Soul.* Oxford: Oxford University Press, 1986.

Taylor, Richard. *Action and Purpose.* Englewood Cliffs, N.J.: Prentice-Hall, 1966.

Schaffer, Campbell, Churchland, and Smith and Jones are good introductions to this area, the latter two leaning heavily toward physicalism.

Index